The Cutting Edge of AI Autonomous Cars

Practical Innovations in
Artificial Intelligence and Machine Learning

Dr. Lance B. Eliot, MBA, PhD

ISBN: 057840964X
ISBN-13: 978-0578409641

DEDICATION

To my incredible daughter, Lauren, and my incredible son, Michael.

Forest fortuna adiuvat (from the Latin; good fortune favors the brave).

CONTENTS

ACKNOWLEDGMENTS

I have been the beneficiary of advice and counsel by many friends, colleagues, family, investors, and many others. I want to thank everyone that has aided me throughout my career. I write from the heart and the head, having experienced first-hand what it means to have others around you that support you during the good times and the tough times.

To Warren Bennis, one of my doctoral advisors and ultimately a colleague, I offer my deepest thanks and appreciation, especially for his calm and insightful wisdom and support.

To Mark Stevens and his generous efforts toward funding and supporting the USC Stevens Center for Innovation.

To Lloyd Greif and the USC Lloyd Greif Center for Entrepreneurial Studies for their ongoing encouragement of founders and entrepreneurs.

To Peter Drucker, William Wang, Aaron Levie, Peter Kim, Jon Kraft, Cindy Crawford, Jenny Ming, Steve Milligan, Chis Underwood, Frank Gehry, Buzz Aldrin, Steve Forbes, Bill Thompson, Dave Dillon, Alan Fuerstman, Larry Ellison, Jim Sinegal, John Sperling, Mark Stevenson, Anand Nallathambi, Thomas Barrack, Jr., and many other innovators and leaders that I have met and gained mightily from doing so.

Thanks to Ed Trainor, Kevin Anderson, James Hickey, Wendell Jones, Ken Harris, DuWayne Peterson, Mike Brown, Jim Thornton, Abhi Beniwal, Al Biland, John Nomura, Eliot Weinman, John Desmond, and many others for their unwavering support during my career.

And most of all thanks as always to Michael and Lauren, for their ongoing support and for having seen me writing and heard much of this material during the many months involved in writing it. To their patience and willingness to listen.

INTRODUCTION

This is a book that provides the newest innovations and the latest Artificial Intelligence (AI) advances about the emerging nature of AI-based autonomous self-driving driverless cars. Via recent advances in Artificial Intelligence (AI) and Machine Learning (ML), we are nearing the day when vehicles can control themselves and will not require and nor rely upon human intervention to perform their driving tasks (or, that allow for human intervention, but only *require* human intervention in very limited ways).

Similar to my other related books, which I describe in a moment and list the chapters in the Appendix A of this book, I am particularly focused on those advances that pertain to self-driving cars. The phrase "autonomous vehicles" is often used to refer to any kind of vehicle, whether it is ground-based or in the air or sea, and whether it is a cargo hauling trailer truck or a conventional passenger car. Though the aspects described in this book are certainly applicable to all kinds of autonomous vehicles, I am focused more so here on cars.

Indeed, I am especially known for my role in aiding the advancement of self-driving cars, serving currently as the Executive Director of the Cybernetic Self-Driving Cars Institute.. In addition to writing software, designing and developing systems and software for self-driving cars, I also speak and write quite a bit about the topic. This book is a collection of some of my more advanced essays. For those of you that might have seen my essays posted elsewhere, I have updated them and integrated them into this book as one handy cohesive package.

You might be interested in companion books that I have written that provide additional key innovations and fundamentals about self-driving cars. Those books are entitled **"Introduction to Driverless Self-Driving Cars," "Advances in AI and Autonomous Vehicles: Cybernetic Self-Driving Cars," "Self-Driving Cars: "The Mother of All AI Projects," "Innovation and Thought Leadership on Self-Driving Driverless Cars," "New Advances in AI Autonomous Driverless Self-Driving Cars,"** and **"Autonomous Vehicle Driverless Self-Driving Cars and**

Artificial Intelligence," "Transformative Artificial Intelligence Driverless Self-Driving Cars," "Disruptive Artificial Intelligence and Driverless Self-Driving Cars, and "State-of-the-Art AI Driverless Self-Driving Cars," and "Top Trends in AI Self-Driving Cars," and "AI Innovations and Self-Driving Cars," "Crucial Advances for AI Driverless Cars," "Sociotechnical Insights and AI Driverless Cars," "Pioneering Advances for AI Driverless Cars" and "Leading Edge Trends for AI Driverless Cars" (they are all available via Amazon). See Appendix A of this herein book to see a listing of the chapters covered in those three books.

For the introduction here to this book, I am going to borrow my introduction from those companion books, since it does a good job of laying out the landscape of self-driving cars and my overall viewpoints on the topic. The remainder of the book is all new material that does not appear in the companion books.

INTRODUCTION TO SELF-DRIVING CARS

This is a book about self-driving cars. Someday in the future, we'll all have self-driving cars and this book will perhaps seem antiquated, but right now, we are at the forefront of the self-driving car wave. Daily news bombards us with flashes of new announcements by one car maker or another and leaves the impression that within the next few weeks or maybe months that the self-driving car will be here. A casual non-technical reader would assume from these news flashes that in fact we must be on the cusp of a true self-driving car.

Here's a real news flash: We are still quite a distance from having a true self-driving car. It is years to go before we get there.

Why is that? Because a true self-driving car is akin to a moonshot. In the same manner that getting us to the moon was an incredible feat, likewise can it be said for achieving a true self-driving car. Anybody that suggests or even brashly states that the true self-driving car is nearly here should be viewed with great skepticism. Indeed, you'll see that I often tend to use the word "hogwash" or "crock" when I assess much of the decidedly *fake news* about self-driving cars. Those of us on the inside know that what is often reported to the outside is malarkey. Few of the insiders are willing to say so. I have no such hesitation.

Indeed, I've been writing a popular blog post about self-driving cars and hitting hard on those that try to wave their hands and pretend that we are on the imminent verge of true self-driving cars. For many years, I've been known as the AI Insider. Besides writing about AI, I also develop AI software. I do what I describe. It also gives me insights into what others that are doing AI are really doing versus what it is said they are doing.

Many faithful readers had asked me to pull together my insightful short essays and put them into another book, which you are now holding in your hands.

For those of you that have been reading my essays over the years, this collection not only puts them together into one handy package, I also updated the essays and added new material. For those of you that are new to the topic of self-driving cars and AI, I hope you find these essays approachable and informative. I also tend to have a writing style with a bit of a voice, and so you'll see that I am times have a wry sense of humor and also like to poke at conformity.

As a former professor and founder of an AI research lab, I for many years wrote in the formal language of academic writing. I published in referred journals and served as an editor for several AI journals. This writing here is not of the nature, and I have adopted a different and more informal style for these essays. That being said, I also do mention from time-to-time more rigorous material on AI and encourage you all to dig into those deeper and more formal materials if so interested.

I am also an AI practitioner. This means that I write AI software for a living. Currently, I head-up the Cybernetics Self-Driving Car Institute, where we are developing AI software for self-driving cars. I am excited to also report that my son, also a software engineer, heads-up our Cybernetics Self-Driving Car Lab. What I have helped to start, and for which he is an integral part, ultimately he will carry long into the future after I have retired. My daughter, a marketing whiz, also is integral to our efforts as head of our Marketing group. She too will carry forward the legacy now being formulated.

For those of you that are reading this book and have a penchant for writing code, you might consider taking a look at the open source code available for self-driving cars. This is a handy place to start learning how to develop AI for self-driving cars. There are also many new educational courses spring forth.

There is a growing body of those wanting to learn about and develop self-driving cars, and a growing body of colleges, labs, and other avenues by which you can learn about self-driving cars.

This book will provide a foundation of aspects that I think will get you ready for those kinds of more advanced training opportunities. If you've already taken those classes, you'll likely find these essays especially interesting as they offer a perspective that I am betting few other instructors or faculty offered to you. These are challenging essays that ask you to think beyond the conventional about self-driving cars.

THE MOTHER OF ALL AI PROJECTS

In June 2017, Apple CEO Tim Cook came out and finally admitted that Apple has been working on a self-driving car. As you'll see in my essays, Apple was enmeshed in secrecy about their self-driving car efforts. We have only been able to read the tea leaves and guess at what Apple has been up to. The notion of an iCar has been floating for quite a while, and self-driving engineers and researchers have been signing tight-lipped Non-Disclosure Agreements (NDA's) to work on projects at Apple that were as shrouded in mystery as any military invasion plans might be.

Tim Cook said something that many others in the Artificial Intelligence (AI) field have been saying, namely, the creation of a self-driving car has got to be the mother of all AI projects. In other words, it is in fact a tremendous moonshot for AI. If a self-driving car can be crafted and the AI works as we hope, it means that we have made incredible strides with AI and that therefore it opens many other worlds of potential breakthrough accomplishments that AI can solve.

Is this hyperbole? Am I just trying to make AI seem like a miracle worker and so provide self-aggrandizing statements for those of us writing the AI software for self-driving cars? No, it is not hyperbole. Developing a true self-driving car is really, really, really hard to do. Let me take a moment to explain why. As a side note, I realize that the Apple CEO is known for at times uttering hyperbole, and he had previously said for example that the year 2012 was "the mother of all years," and he had said that the release of iOS 10 was "the mother of all releases" – all of which does suggest he likes to use the handy "mother of" expression. But, I assure you, in terms of true self-driving cars, he has hit the nail on the head. For sure.

When you think about a moonshot and how we got to the moon, there are some identifiable characteristics and those same aspects can be applied to creating a true self-driving car. You'll notice that I keep putting the word "true" in front of the self-driving car expression. I do so because as per my essay about the various levels of self-driving cars, there are some self-driving cars that are only somewhat of a self-driving car. The somewhat versions are ones that require a human driver to be ready to intervene. In my view, that's not a true self-driving car. A true self-driving car is one that requires no human driver intervention at all. It is a car that can entirely undertake via automation the driving task without any human driver needed. This is the essence of what is known as a Level 5 self-driving car. We are currently at the Level 2 and Level 3 mark, and not yet at Level 5.

Getting to the moon involved aspects such as having big stretch goals, incremental progress, experimentation, innovation, and so on. Let's review how this applied to the moonshot of the bygone era, and how it applies to the self-driving car moonshot of today.

Big Stretch Goal

Trying to take a human and deliver the human to the moon, and bring them back, safely, was an extremely large stretch goal at the time. No one knew whether it could be done. The technology wasn't available yet. The cost was huge. The determination would need to be fierce. Etc. To reach a Level 5 self-driving car is going to be the same. It is a big stretch goal. We can readily get to the Level 3, and we are able to see the Level 4 just up ahead, but a Level 5 is still an unknown as to if it is doable. It should eventually be doable and in the same way that we thought we'd eventually get to the moon, but when it will occur is a different story.

Incremental Progress

Getting to the moon did not happen overnight in one fell swoop. It took years and years of incremental progress to get there. Likewise for self-driving cars. Google has famously been striving to get to the Level 5, and pretty much been willing to forgo dealing with the intervening levels, but most of the other self-driving car makers are doing the incremental route. Let's get a good Level 2 and a somewhat Level 3 going. Then, let's improve the Level 3 and get a somewhat Level 4 going. Then, let's improve the Level 4 and finally arrive at a Level 5. This seems to be the prevalent way that we are going to achieve the true self-driving car.

Experimentation

You likely know that there were various experiments involved in perfecting the approach and technology to get to the moon. As per making incremental progress, we first tried to see if we could get a rocket to go into space and safety return, then put a monkey in there, then with a human, then we went all the way to the moon but didn't land, and finally we arrived at the mission that actually landed on the moon. Self-driving cars are the same way. We are doing simulations of self-driving cars. We do testing of self-driving cars on private land under controlled situations. We do testing of self-driving cars on public roadways, often having to meet regulatory requirements including for example having an engineer or equivalent in the car to take over the controls if needed. And so on. Experiments big and small are needed to figure out what works and what doesn't.

Innovation

There are already some advances in AI that are allowing us to progress toward self-driving cars. We are going to need even more advances. Innovation in all aspects of technology are going to be required to achieve a true self-driving car. By no means do we already have everything in-hand that we need to get there. Expect new inventions and new approaches, new algorithms, etc.

Setbacks

Most of the pundits are avoiding talking about potential setbacks in the progress toward self-driving cars. Getting to the moon involved many setbacks, some of which you never have heard of and were buried at the time so as to not dampen enthusiasm and funding for getting to the moon. A recurring theme in many of my included essays is that there are going to be setbacks as we try to arrive at a true self-driving car. Take a deep breath and be ready. I just hope the setbacks don't completely stop progress. I am sure that it will cause progress to alter in a manner that we've not yet seen in the self-driving car field. I liken the self-driving car of today to the excitement everyone had for Uber when it first got going. Today, we have a different view of Uber and with each passing day there are more regulations to the ride sharing business and more concerns raised. The darling child only stays a darling until finally that child acts up. It will happen the same with self-driving cars.

SELF-DRIVING CARS CHALLENGES

But what exactly makes things so hard to have a true self-driving car, you might be asking. You have seen cruise control for years and years. You've lately seen cars that can do parallel parking. You've seen YouTube videos of Tesla drivers that put their hands out the window as their car zooms along the highway, and seen to therefore be in a self-driving car. Aren't we just needing to put a few more sensors onto a car and then we'll have in-hand a true self-driving car? Nope.

Consider for a moment the nature of the driving task. We don't just let anyone at any age drive a car. Worldwide, most countries won't license a driver until the age of 18, though many do allow a learner's permit at the age of 15 or 16. Some suggest that a younger age would be physically too small to reach the controls of the car. Though this might be the case, we could easily adjust the controls to allow for younger aged and thus smaller stature. It's not their physical size that matters. It's their cognitive development that matters.

To drive a car, you need to be able to reason about the car, what the car can and cannot do. You need to know how to operate the car. You need to know about how other cars on the road drive. You need to know what is allowed in driving such as speed limits and driving within marked lanes. You need to be able to react to situations and be able to avoid getting into accidents. You need to ascertain when to hit your brakes, when to steer clear of a pedestrian, and how to keep from ramming that motorcyclist that just cut you off.

Many of us had taken courses on driving. We studied about driving and took driver training. We had to take a test and pass it to be able to drive. The point being that though most adults take the driving task for granted, and we often "mindlessly" drive our cars, there is a significant amount of cognitive effort that goes into driving a car. After a while, it becomes second nature. You don't especially think about how you drive, you just do it. But, if you watch a novice driver, say a teenager learning to drive, you suddenly realize that there is a lot more complexity to it than we seem to realize.

Furthermore, driving is a very serious task. I recall when my daughter and son first learned to drive. They are both very conscientious people. They wanted to make sure that whatever they did, they did well, and that they did not harm anyone. Every day, when you get into a car, it is probably around 4,000 pounds of hefty metal and plastics (about two tons), and it is a lethal weapon. Think about it. You drive down the street in an object that weighs two tons and with the engine it can accelerate and ram into anything you want to hit. The damage a car can inflict is very scary. Both my children were surprised that they were being given the right to maneuver this monster of a beast that could cause tremendous harm entirely by merely letting go of the steering wheel for a moment or taking your eyes off the road.

In fact, in the United States alone there are about 30,000 deaths per year by auto accidents, which is around 100 per day. Given that there are about 263 million cars in the United States, I am actually more amazed that the number of fatalities is not a lot higher. During my morning commute, I look at all the thousands of cars on the freeway around me, and I think that if all of them decided to go zombie and drive in a crazy maniac way, there would be many people dead. Somehow, incredibly, each day, most people drive relatively safely. To me, that's a miracle right there. Getting millions and millions of people to be safe and sane when behind the wheel of a two ton mobile object, it's a feat that we as a society should admire with pride.

So, hopefully you are in agreement that the driving task requires a great deal of cognition. You don't' need to be especially smart to drive a car, and we've done quite a bit to make car driving viable for even the average dolt. There isn't an IQ test that you need to take to drive a car. If you can read and write, and pass a test, you pretty much can legally drive a car. There are of course some that drive a car and are not legally permitted to do so, plus there are private areas such as farms where drivers are young, but for public roadways in the United States, you can be generally of average intelligence (or less) and be able to legally drive.

This though makes it seem like the cognitive effort must not be much. If the cognitive effort was truly hard, wouldn't we only have Einstein's that could drive a car? We have made sure to keep the driving task as simple as we can, by making the controls easy and relatively standardized, and by having roads that are relatively standardized, and so on. It is as though Disneyland has put their Autopia into the real-world, by us all as a society agreeing that roads will be a certain way, and we'll all abide by the various rules of driving.

A modest cognitive task by a human is still something that stymies AI. You certainly know that AI has been able to beat chess players and be good at other kinds of games. This type of narrow cognition is not what car driving is about. Car driving is much wider. It requires knowledge about the world, which a chess playing AI system does not need to know. The cognitive aspects of driving are on the one hand seemingly simple, but at the same time require layer upon layer of knowledge about cars, people, roads, rules, and a myriad of other "common sense" aspects. We don't have any AI systems today that have that same kind of breadth and depth of awareness and knowledge.

As revealed in my essays, the self-driving car of today is using trickery to do particular tasks. It is all very narrow in operation. Plus, it currently assumes that a human driver is ready to intervene. It is like a child that we have taught to stack blocks, but we are needed to be right there in case the child stacks them too high and they begin to fall over. AI of today is brittle, it is narrow, and it does not approach the cognitive abilities of humans. This is why the true self-driving car is somewhere out in the future.

Another aspect to the driving task is that it is not solely a mind exercise. You do need to use your senses to drive. You use your eyes a vision sensors to see the road ahead. You vision capability is like a streaming video, which your brain needs to continually analyze as you drive. Where is the road? Is there a pedestrian in the way? Is there another car ahead of you? Your senses are relying a flood of info to your brain. Self-driving cars are trying to do the same, by using cameras, radar, ultrasound, and lasers. This is an attempt at mimicking how humans have senses and sensory apparatus.

Thus, the driving task is mental and physical. You use your senses, you use your arms and legs to manipulate the controls of the car, and you use your brain to assess the sensory info and direct your limbs to act upon the controls of the car. This all happens instantly. If you've ever perhaps gotten something in your eye and only had one eye available to drive with, you suddenly realize how dependent upon vision you are. If you have a broken foot with a cast, you suddenly realize how hard it is to control the brake pedal and the accelerator. If you've taken medication and your brain is maybe sluggish, you suddenly realize how much mental strain is required to drive a car.

An AI system that plays chess only needs to be focused on playing chess. The physical aspects aren't important because usually a human moves the chess pieces or the chessboard is shown on an electronic display. Using AI for a more life-and-death task such as analyzing MRI images of patients, this again does not require physical capabilities and instead is done by examining images of bits.

Driving a car is a true life-and-death task. It is a use of AI that can easily and at any moment produce death. For those colleagues of mine that are developing this AI, as am I, we need to keep in mind the somber aspects of this. We are producing software that will have in its virtual hands the lives of the occupants of the car, and the lives of those in other nearby cars, and the lives of nearby pedestrians, etc. Chess is not usually a life-or-death matter.

Driving is all around us. Cars are everywhere. Most of today's AI applications involve only a small number of people. Or, they are behind the scenes and we as humans have other recourse if the AI messes up. AI that is driving a car at 80 miles per hour on a highway had better not mess up. The consequences are grave. Multiply this by the number of cars, if we could put magically self-driving into every car in the USA, we'd have AI running in the 263 million cars. That's a lot of AI spread around. This is AI on a massive scale that we are not doing today and that offers both promise and potential peril.

There are some that want AI for self-driving cars because they envision a world without any car accidents. They envision a world in which there is no car congestion and all cars cooperate with each other. These are wonderful utopian visions.

They are also very misleading. The adoption of self-driving cars is going to be incremental and not overnight. We cannot economically just junk all existing cars. Nor are we going to be able to affordably retrofit existing cars. It is more likely that self-driving cars will be built into new cars and that over many years of gradual replacement of existing cars that we'll see the mix of self-driving cars become substantial in the real-world.

In these essays, I have tried to offer technological insights without being overly technical in my description, and also blended the business, societal, and economic aspects too. Technologists need to consider the nontechnological impacts of what they do. Non-technologists should be aware of what is being developed.

We all need to work together to collectively be prepared for the enormous disruption and transformative aspects of true self-driving cars. We all need to be involved in this mother of all AI projects.

WHAT THIS BOOK PROVIDES

What does this book provide to you? It introduces many of the key elements about self-driving cars and does so with an AI based perspective. I weave together technical and non-technical aspects, readily going from being concerned about the cognitive capabilities of the driving task and how the technology is embodying this into self-driving cars, and in the next breath I discuss the societal and economic aspects.

They are all intertwined because that's the way reality is. You cannot separate out the technology per se, and instead must consider it within the milieu of what is being invented and innovated, and do so with a mindset towards the contemporary mores and culture that shape what we are doing and what we hope to do.

WHY THIS BOOK

I wrote this book to try and bring to the public view many aspects about self-driving cars that nobody seems to be discussing.

For business leaders that are either involved in making self-driving cars or that are going to leverage self-driving cars, I hope that this book will enlighten you as to the risks involved and ways in which you should be strategizing about how to deal with those risks.

For entrepreneurs, startups and other businesses that want to enter into the self-driving car market that is emerging, I hope this book sparks your interest in doing so, and provides some sense of what might be prudent to pursue.

For researchers that study self-driving cars, I hope this book spurs your interest in the risks and safety issues of self-driving cars, and also nudges you toward conducting research on those aspects.

For students in computer science or related disciplines, I hope this book will provide you with interesting and new ideas and material, for which you might conduct research or provide some career direction insights for you.

For AI companies and high-tech companies pursuing self-driving cars, this book will hopefully broaden your view beyond just the mere coding and development needed to make self-driving cars.

For all readers, I hope that you will find the material in this book to be stimulating. Some of it will be repetitive of things you already know. But I am pretty sure that you'll also find various eureka moments whereby you'll discover a new technique or approach that you had not earlier thought of. I am also betting that there will be material that forces you to rethink some of your current practices.

I am not saying you will suddenly have an epiphany and change what you are doing. I do think though that you will reconsider or perhaps revisit what you are doing.

For anyone choosing to use this book for teaching purposes, please take a look at my suggestions for doing so, as described in the Appendix. I have found the material handy in courses that I have taught, and likewise other faculty have told me that they have found the material handy, in some cases as extended readings and in other instances as a core part of their course (depending on the nature of the class).

In my writing for this book, I have tried carefully to blend both the practitioner and the academic styles of writing. It is not as dense as is typical academic journal writing, but at the same time offers depth by going into the nuances and trade-offs of various practices.

The word "deep" is in vogue today, meaning getting deeply into a subject or topic, and so is the word "unpack" which means to tease out the underlying aspects of a subject or topic. I have sought to offer material that addresses an issue or topic by going relatively deeply into it and make sure that it is well unpacked.

Finally, in any book about AI, it is difficult to use our everyday words without having some of them be misinterpreted. Specifically, it is easy to anthropomorphize AI. When I say that an AI system "knows" something, I do not want you to construe that the AI system has sentience and "knows" in the same way that humans do. They aren't that way, as yet. I have tried to use quotes around such words from time-to-time to emphasize that the words I am using should not be misinterpreted to ascribe true human intelligence to the AI systems that we know of today. If I used quotes around all such words, the book would be very difficult to read, and so I am doing so judiciously. Please keep that in mind as you read the material, thanks.

COMPANION BOOKS

If you find this material of interest, you might want to also see my other books on self-driving cars, entitled:

1. **"Introduction to Driverless Self-Driving Cars"** by Dr. Lance Eliot

2. **"Innovation and Thought Leadership on Self-Driving Driverless Cars"** by Dr. Lance Eliot

3. **"Advances in AI and Autonomous Vehicles: Cybernetic Self-Driving Cars"** by Dr. Lance Eliot

4. *"Self-Driving Cars: The Mother of All AI Projects"* by Dr. Lance Eliot

5. **"New Advances in AI Autonomous Driverless Self-Driving Cars"** by Dr. Lance Eliot

6. **"Autonomous Vehicle Driverless Self-Driving Cars and Artificial Intelligence"** by Dr. Lance Eliot and Michael B. Eliot

7. **"Transformative Artificial Intelligence Driverless Self-Driving Cars"** by Dr. Lance Eliot

8. **"Disruptive Artificial Intelligence and Driverless Self-Driving Cars"** by Dr. Lance Eliot

9. "State-of-the-Art AI Driverless Self-Driving Cars" by Dr. Lance Eliot

10. "Top Trends in AI Self-Driving Cars" by Dr. Lance Eliot

11. **"AI Innovations and Self-Driving Cars"** by Dr. Lance Eliot

12. **"Crucial Advances for AI Driverless Cars"** by Dr. Lance Eliot

13. **"Sociotechnical Insights and AI Driverless Cars"** by Dr. Lance Eliot.

14. **"Pioneering Advances for AI Driverless Cars"** by Dr. Lance Eliot

15. **"Leading Edge Trends for AI Driverless Cars"** by Dr. Lance Eliot

16. **"The Cutting Edge of AI Autonomous Cars"** by Dr. Lance Eliot

All of the above books are available on Amazon and at other major global booksellers.

CHAPTER 1

ELIOT FRAMEWORK FOR AI SELF-DRIVING CARS

CHAPTER 1

ELIOT FRAMEWORK FOR AI SELF-DRIVING CARS

This chapter is a core foundational aspect for understanding AI self-driving cars and I have used this same chapter in several of my other books to introduce the reader to essential elements of this field. Once you've read this chapter, you'll be prepared to read the rest of the material since the foundational essence of the components of autonomous AI driverless self-driving cars will have been established for you.

———————

When I give presentations about self-driving cars and teach classes on the topic, I have found it helpful to provide a framework around which the various key elements of self-driving cars can be understood and organized (see diagram at the end of this chapter). The framework needs to be simple enough to convey the overarching elements, but at the same time not so simple that it belies the true complexity of self-driving cars. As such, I am going to describe the framework here and try to offer in a thousand words (or more!) what the framework diagram itself intends to portray.

The core elements on the diagram are numbered for ease of reference. The numbering does not suggest any kind of prioritization of the elements. Each element is crucial. Each element has a purpose, and otherwise would not be included in the framework. For some self-driving cars, a particular element might be more important or somehow distinguished in comparison to other self-driving cars.

You could even use the framework to rate a particular self-driving car, doing so by gauging how well it performs in each of the elements of the framework. I will describe each of the elements, one at a time. After doing so, I'll discuss aspects that illustrate how the elements interact and perform during the overall effort of a self-driving car.

At the Cybernetic Self-Driving Car Institute, we use the framework to keep track of what we are working on, and how we are developing software that fills in what is needed to achieve Level 5 self-driving cars.

D-01: Sensor Capture

Let's start with the one element that often gets the most attention in the press about self-driving cars, namely, the sensory devices for a self-driving car.

On the framework, the box labeled as D-01 indicates "Sensor Capture" and refers to the processes of the self-driving car that involve collecting data from the myriad of sensors that are used for a self-driving car. The types of devices typically involved are listed, such as the use of mono cameras, stereo cameras, LIDAR devices, radar systems, ultrasonic devices, GPS, IMU, and so on.

These devices are tasked with obtaining data about the status of the self-driving car and the world around it. Some of the devices are continually providing updates, while others of the devices await an indication by the self-driving car that the device is supposed to collect data. The data might be first transformed in some fashion by the device itself, or it might instead be fed directly into the sensor capture as raw data. At that point, it might be up to the sensor capture processes to do transformations on the data. This all varies depending upon the nature of the devices being used and how the devices were designed and developed.

D-02: Sensor Fusion

Imagine that your eyeballs receive visual images, your nose receives odors, your ears receive sounds, and in essence each of your distinct sensory devices is getting some form of input. The input befits the nature of the device. Likewise, for a self-driving car, the cameras provide visual images, the radar returns radar reflections, and so on.

Each device provides the data as befits what the device does.

At some point, using the analogy to humans, you need to merge together what your eyes see, what your nose smells, what your ears hear, and piece it all together into a larger sense of what the world is all about and what is happening around you. Sensor fusion is the action of taking the singular aspects from each of the devices and putting them together into a larger puzzle.

Sensor fusion is a tough task. There are some devices that might not be working at the time of the sensor capture. Or, there might some devices that are unable to report well what they have detected. Again, using a human analogy, suppose you are in a dark room and so your eyes cannot see much. At that point, you might need to rely more so on your ears and what you hear. The same is true for a self-driving car. If the cameras are obscured due to snow and sleet, it might be that the radar can provide a greater indication of what the external conditions consist of.

In the case of a self-driving car, there can be a plethora of such sensory devices. Each is reporting what it can. Each might have its difficulties. Each might have its limitations, such as how far ahead it can detect an object. All of these limitations need to be considered during the sensor fusion task.

D-03: Virtual World Model

For humans, we presumably keep in our minds a model of the world around us when we are driving a car. In your mind, you know that the car is going at say 60 miles per hour and that you are on a freeway. You have a model in your mind that your car is surrounded by other cars, and that there are lanes to the freeway. Your model is not only based on what you can see, hear, etc., but also what you know about the nature of the world. You know that at any moment that car ahead of you can smash on its brakes, or the car behind you can ram into your car, or that the truck in the next lane might swerve into your lane.

The AI of the self-driving car needs to have a virtual world model, which it then keeps updated with whatever it is receiving from the sensor fusion, which received its input from the sensor capture and the sensory devices.

D-04: System Action Plan

By having a virtual world model, the AI of the self-driving car is able to keep track of where the car is and what is happening around the car. In addition, the AI needs to determine what to do next. Should the self-driving car hit its brakes? Should the self-driving car stay in its lane or swerve into the lane to the left? Should the self-driving car accelerate or slow down?

A system action plan needs to be prepared by the AI of the self-driving car. The action plan specifies what actions should be taken. The actions need to pertain to the status of the virtual world model. Plus, the actions need to be realizable.

This realizability means that the AI cannot just assert that the self-driving car should suddenly sprout wings and fly. Instead, the AI must be bound by whatever the self-driving car can actually do, such as coming to a halt in a distance of X feet at a speed of Y miles per hour, rather than perhaps asserting that the self-driving car come to a halt in 0 feet as though it could instantaneously come to a stop while it is in motion.

D-05: Controls Activation

The system action plan is implemented by activating the controls of the car to act according to what the plan stipulates. This might mean that the accelerator control is commanded to increase the speed of the car. Or, the steering control is commanded to turn the steering wheel 30 degrees to the left or right.

One question arises as to whether or not the controls respond as they are commanded to do. In other words, suppose the AI has commanded the accelerator to increase, but for some reason it does not do so. Or, maybe it tries to do so, but the speed of the car does not increase. The controls activation feeds back into the virtual world model, and simultaneously the virtual world model is getting updated from the sensors, the sensor capture, and the sensor fusion. This allows the AI to ascertain what has taken place as a result of the controls being commanded to take some kind of action.

By the way, please keep in mind that though the diagram seems to have a linear progression to it, the reality is that these are all aspects of the self-driving car that are happening in parallel and simultaneously. The sensors are capturing data, meanwhile the sensor fusion is taking place, meanwhile the virtual model is being updated, meanwhile the system action plan is being formulated and reformulated, meanwhile the controls are being activated.

This is the same as a human being that is driving a car. They are eyeballing the road, meanwhile they are fusing in their mind the sights, sounds, etc., meanwhile their mind is updating their model of the world around them, meanwhile they are formulating an action plan of what to do, and meanwhile they are pushing their foot onto the pedals and steering the car. In the normal course of driving a car, you are doing all of these at once. I mention this so that when you look at the diagram, you will think of the boxes as processes that are all happening at the same time, and not as though only one happens and then the next.

They are shown diagrammatically in a simplistic manner to help comprehend what is taking place. You though should also realize that they are working in parallel and simultaneous with each other. This is a tough aspect in that the inter-element communications involve latency and other aspects that must be taken into account. There can be delays in one element updating and then sharing its latest status with other elements.

D-06: Automobile & CAN

Contemporary cars use various automotive electronics and a Controller Area Network (CAN) to serve as the components that underlie the driving aspects of a car. There are Electronic Control Units (ECU's) which control subsystems of the car, such as the engine, the brakes, the doors, the windows, and so on.

The elements D-01, D-02, D-03, D-04, D-05 are layered on top of the D-06, and must be aware of the nature of what the D-06 is able to do and not do.

D-07: In-Car Commands

Humans are going to be occupants in self-driving cars. In a Level 5 self-driving car, there must be some form of communication that takes place between the humans and the self-driving car. For example, I go into a self-driving car and tell it that I want to be driven over to Disneyland, and along the way I want to stop at In-and-Out Burger. The self-driving car now parses what I've said and tries to then establish a means to carry out my wishes.

In-car commands can happen at any time during a driving journey. Though my example was about an in-car command when I first got into my self-driving car, it could be that while the self-driving car is carrying out the journey that I change my mind. Perhaps after getting stuck in traffic, I tell the self-driving car to forget about getting the burgers and just head straight over to the theme park. The self-driving car needs to be alert to in-car commands throughout the journey.

D-08: VX2 Communications

We will ultimately have self-driving cars communicating with each other, doing so via V2V (Vehicle-to-Vehicle) communications. We will also have self-driving cars that communicate with the roadways and other aspects of the transportation infrastructure, doing so via V2I (Vehicle-to-Infrastructure).

The variety of ways in which a self-driving car will be communicating with other cars and infrastructure is being called V2X, whereby the letter X means whatever else we identify as something that a car should or would want to communicate with. The V2X communications will be taking place simultaneous with everything else on the diagram, and those other elements will need to incorporate whatever it gleans from those V2X communications.

D-09: Deep Learning

The use of Deep Learning permeates all other aspects of the self-driving car. The AI of the self-driving car will be using deep learning to do a better job at the systems action plan, and at the controls activation, and at the sensor fusion, and so on.

Currently, the use of artificial neural networks is the most prevalent form of deep learning. Based on large swaths of data, the neural networks attempt to "learn" from the data and therefore direct the efforts of the self-driving car accordingly.

D-10: Tactical AI

Tactical AI is the element of dealing with the moment-to-moment driving of the self-driving car. Is the self-driving car staying in its lane of the freeway? Is the car responding appropriately to the controls commands? Are the sensory devices working?

For human drivers, the tactical equivalent can be seen when you watch a novice driver such as a teenager that is first driving. They are focused on the mechanics of the driving task, keeping their eye on the road while also trying to properly control the car.

D-11: Strategic AI

The Strategic AI aspects of a self-driving car are dealing with the larger picture of what the self-driving car is trying to do. If I had asked that the self-driving car take me to Disneyland, there is an overall journey map that needs to be kept and maintained.

There is an interaction between the Strategic AI and the Tactical AI. The Strategic AI is wanting to keep on the mission of the driving, while the Tactical AI is focused on the particulars underway in the driving effort. If the Tactical AI seems to wander away from the overarching mission, the Strategic AI wants to see why and get things back on track. If the Tactical AI realizes that there is something amiss on the self-driving car, it needs to alert the Strategic AI accordingly and have an adjustment to the overarching mission that is underway.

D-12: Self-Aware AI

Very few of the self-driving cars being developed are including a Self-Aware AI element, which we at the Cybernetic Self-Driving Car Institute believe is crucial to Level 5 self-driving cars.

The Self-Aware AI element is intended to watch over itself, in the sense that the AI is making sure that the AI is working as intended. Suppose you had a human driving a car, and they were starting to drive erratically. Hopefully, their own self-awareness would make them realize they themselves are driving poorly, such as perhaps starting to fall asleep after having been driving for hours on end. If you had a passenger in the car, they might be able to alert the driver if the driver is starting to do something amiss. This is exactly what the Self-Aware

AI element tries to do, it becomes the overseer of the AI, and tries to detect when the AI has become faulty or confused, and then find ways to overcome the issue.

D-13: Economic

The economic aspects of a self-driving car are not per se a technology aspect of a self-driving car, but the economics do indeed impact the nature of a self-driving car. For example, the cost of outfitting a self-driving car with every kind of possible sensory device is prohibitive, and so choices need to be made about which devices are used. And, for those sensory devices chosen, whether they would have a full set of features or a more limited set of features.

We are going to have self-driving cars that are at the low-end of a consumer cost point, and others at the high-end of a consumer cost point. You cannot expect that the self-driving car at the low-end is going to be as robust as the one at the high-end. I realize that many of the self-driving car pundits are acting as though all self-driving cars will be the same, but they won't be. Just like anything else, we are going to have self-driving cars that have a range of capabilities. Some will be better than others. Some will be safer than others. This is the way of the real-world, and so we need to be thinking about the economics aspects when considering the nature of self-driving cars.

D-14: Societal

This component encompasses the societal aspects of AI which also impacts the technology of self-driving cars. For example, the famous Trolley Problem involves what choices should a self-driving car make when faced with life-and-death matters. If the self-driving car is about to either hit a child standing in the roadway, or instead ram into a tree at the side of the road and possibly kill the humans in the self-driving car, which choice should be made?

We need to keep in mind the societal aspects will underlie the AI of the self-driving car. Whether we are aware of it explicitly or not, the AI will have embedded into it various societal assumptions.

D-15: Innovation

I included the notion of innovation into the framework because we can anticipate that whatever a self-driving car consists of, it will continue to be innovated over time. The self-driving cars coming out in the next several years will undoubtedly be different and less innovative than the versions that come out in ten years hence, and so on.

Framework Overall

For those of you that want to learn about self-driving cars, you can potentially pick a particular element and become specialized in that aspect. Some engineers are focusing on the sensory devices. Some engineers focus on the controls activation. And so on. There are specialties in each of the elements.

Researchers are likewise specializing in various aspects. For example, there are researchers that are using Deep Learning to see how best it can be used for sensor fusion. There are other researchers that are using Deep Learning to derive good System Action Plans. Some are studying how to develop AI for the Strategic aspects of the driving task, while others are focused on the Tactical aspects.

A well-prepared all-around software developer that is involved in self-driving cars should be familiar with all of the elements, at least to the degree that they know what each element does. This is important since whatever piece of the pie that the software developer works on, they need to be knowledgeable about what the other elements are doing.

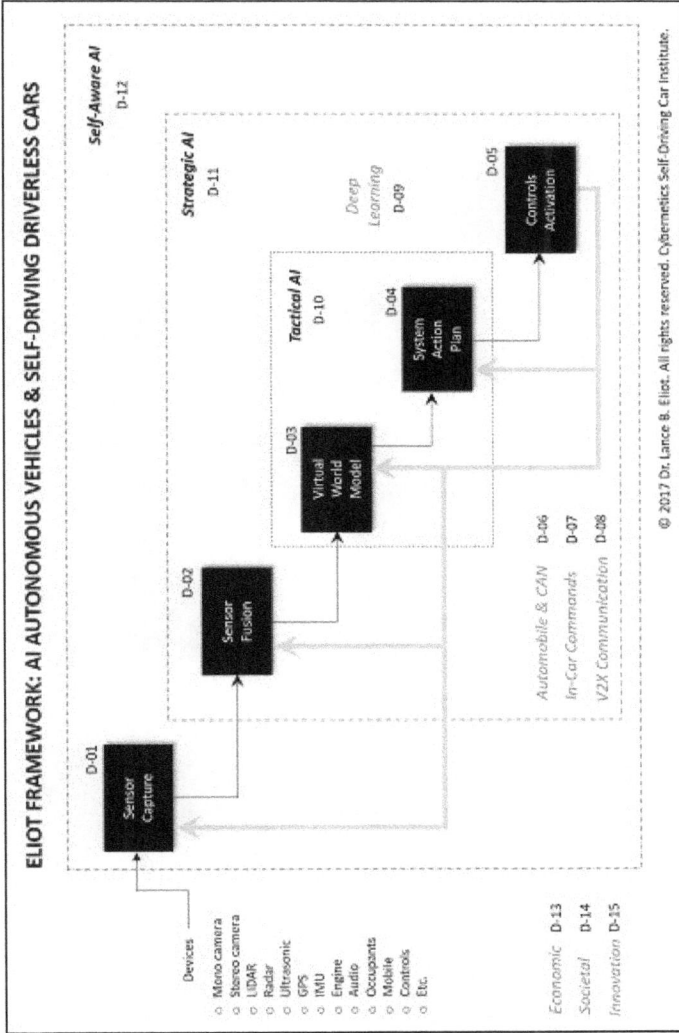

ELIOT FRAMEWORK: AI AUTONOMOUS VEHICLES & SELF-DRIVING DRIVERLESS CARS

Self-Aware AI D-12

Strategic AI D-11

Deep Learning D-09

Tactical AI D-10

Sensor Capture D-01

Sensor Fusion D-02

Virtual World Model D-03

System Action Plan D-04

Controls Activation D-05

Devices
- Mono camera
- Stereo camera
- LIDAR
- Radar
- Ultrasonic
- GPS
- IMU
- Engine
- Audio
- Occupants
- Mobile
- Controls
- Etc.

Automobile & CAN D-06
In-Car Commands D-07
V2X Communication D-08

Economic D-13
Societal D-14
Innovation D-15

CHAPTER 2

DRIVING CONTROLS AND AI SELF-DRIVING CARS

CHAPTER 2

DRIVING CONTROLS AND AI SELF-DRIVING CARS

What's the deal about the driver controls in AI self-driving cars?

That's one of the most popular questions I get asked when I am presenting at AI self-driving car events and Autonomous Vehicles (AV) conferences.

At the Cybernetic AI Self-Driving Car Institute, we are developing AI software for self-driving cars, and the aspects of driver controls are also of crucial attention to our efforts, along with being notable for the efforts of the auto makers and other tech firms that are developing self-driving cars or so-called driverless or robot cars.

If you are willing to strap-in and put on your seat belt, I'll do a whirlwind tour through the nuances of the ongoing debate about driver car controls in AI self-driving cars. It's quite a story and it has both up's and down's, which might leave you in tears or you might be uplifted. We'll see.

In essence, the matter deals with whether or not there should be a steering wheel, a brake pedal, and an accelerator pedal -- which I'll henceforth herein refer to collectively as "driver controls," provided in AI self-driving cars.

When I say the phrase "driver controls" please be aware that I am really referring to "human driver controls" and I'm going to leave out the word "human" for brevity sake. Nonetheless, I trust that you will realize that my utterances of "driver controls" suggests a set of human usable means to steer, brake, and accelerate a car, of which the most common instantiation consists of the conventional steering wheel, brake pedal, and accelerator that we all have come to know and love (well, maybe we like it, or maybe we are just used to those mechanisms as a convenient means of being able to direct the motions of a car, and thus by rote familiarity we have accepted it as appropriate and the right way to drive!).

I'd like to unpack this driver controls topic and will reveal to you that there is more to the matter than perhaps meets the eye.

First, look off into the future.

If you were to take a close gander at the various concept cars of the future, you'll notice that by-and-large they are depicted as having no apparent driver controls. Indeed, we've all grown-up with animated cartoons that show autonomous self-driving cars and they don't have any conventional car driving controls portrayed. Just as we have been promised jet packs that will allow us to fly around however we want, it seems too that we have been promised there will be cars without any kind of driver control contraptions.

On the surface, this seems to make sense. Why not leave the driving to the AI? It is handy to let someone else, or shall we say something else, deal with the arduous chore of driving a car. Logically, if the AI is going to be doing the driving, we can therefore assume the human is not going to be doing the driving, and therefore we can eliminate the usual driver controls consisting of the steering wheel and pedals. This all seems quite sensible and logical.

It is kind of nifty to be able to potentially get rid of those pesky driver controls. By doing so, you open up new possibilities of a major redesign for the interior of cars. Currently, suppose you had to design a car. Your immediate and obvious constraint is that you need to put towards the front of the interior a set of driver controls. It needs to be at the front of the interior so that the human driver can readily see out of the car and look at the road ahead.

You can potentially place the driver controls on the left side or the right side of the interior, though this too is constrained by the country or region that you are aiming to have your car used in. There needs to be a seat for the driver, thus you have another constraint in that you are now allocating interior space at the front that is either to the right or left and that has to have a seat. I suppose you could decide that the driver will stand-up or be laying down while they drive, but this isn't particularly acceptable for today's cars.

So far, in your design of a car for today's needs, you've already been forced to set aside perhaps one-fourth or one-fifth of the interior, doing so for a four-seater sized car. You can then try to play around with the rest of the interior design, but it is going to be hard because you've already got that albatross hanging around your neck of the pre-determined space consumed for the driver controls and the seating for the human driver.

I've seen some concept cars that have the human driver wearing a Virtual Reality (VR) headset, and thus this presumably allows the car designer to put the driver anywhere in the car. In essence, you could become a true backseat driver by donning a VR headset and having the driver controls placed at the back of the front seats of the car. Frankly, I doubt that we'll be seeing cars of that ilk. I'd bet that we'll be experiencing AI self-driving cars before there becomes any outcry to have VR headset driving cars by humans. I'm willing to put $5 on that wager.

Overall, the point being that the use of conventional driving controls has a tendency to dictate what the interior design of a car must consist of. If you could eliminate the driver controls, you now would have an evergreen slate to be able to use the entire interior of the car in whatever way you fancied. Maybe put swiveling passenger seats and let the occupants turn in whatever direction they like. Maybe put beds into the car and let people laydown and catch some shut-eye while in the car. You name it.

Overall, if we could somehow omit the driver controls, it would allow for a re-imagining of the interior of cars. Before we completely jump on board that bandwagon, I'll mention a somewhat quirky aside on the matter.

Shift your mindset somewhat beyond the topic of cars per se.

Suppose that we are able to someday (soon?) build robots that look like humans and have sufficient AI to perform human-like tasks, including that these robots would have legs, arms, feet, hands, heads, etc. that allows them to undertake such human-like tasks. We see this in science fiction movies quite frequently, though for now I'd like you to think of these real-world futuristic robots as benevolent and not aiming to destroy mankind. It seems that most science fiction portrayals usually have the humanoid-like robots opting to wipe out humanity. I'm not going to tackle that aspect herein and just go along with the assumption that these are robots which are truly for the benefit of us humans.

If we had robots that could walk and talk like us and had AI that was sufficiently skilled such that they could drive a car, we might then be able to keep cars designed as they are today and retain the driver controls. In essence, rather than trying to switch out today's cars with the upcoming new-fangled AI self-driving cars, we could instead perhaps cruise along with today's cars and still have them being driven in a "driverless" way (well, a "human driverless" way, via the robots that walk and talk).

I bring this up because today we have around 250+ million conventional cars in the United States alone. Those conventional cars are not going to overnight transform into AI self-driving cars. In fact, it is unlikely that those conventional cars can somehow be retrofitted into becoming AI self-driving cars. Overall, we'll need to buy new cars, ones that are AI self-driving cars. What then happens to the millions upon millions of older conventional cars?

The odds are that those conventional cars will remain around for a very long time. I mention this because there are some pundits that keep referring to a Utopian world in which there are only AI self-driving cars on our public roads. I doubt this will happen for a very, very, very long time (if at all).

Indeed, the use of human driven cars will last for many years, likely many decades, and the advent of AI self-driving cars will occur while there are still human driven cars on the roads. This is a crucial idea since this means that the AI of self-driving cars needs to be able to contend with not just other AI self-driving cars, but also contend with human driven cars. It is easy to envision a simplistic and rather unrealistic world in which all AI self-driving cars are politely interacting with each other and being civil about roadway interactions. That's not what is going to be happening for the foreseeable future. AI self-driving cars and human driven cars will need to be able to cope with each other. Period.

In any case, the idea is that if we are able to develop walking and talking robots, they presumably could then possibly act as drivers of cars, and since they are shaped the same way we humans are, they could sit in the driver's seat and work the driver controls for us.

I went laterally on this tangent herein and had warned you that it would be a somewhat quirky angle on this topic.

It is admittedly a bit unnerving to consider getting into your personal car as a front-seat passenger and having your robot chauffeur that gets into the car with you, sitting down at the car controls and asking you where you want to go. You then sit in your passenger seat as the robot drives, and observe it looking forward, turning its robotic head as needed, and moving its robotic arms and legs to manipulate the steering wheel and the pedals. Just hope that the robot driver doesn't opt to calculate pi or start yammering away on its smartphone and become a distracted driver (ha!). Plus, do you think it knows to click-it or ticket it (i.e., the National Highway Traffic Safety Administration's national campaign for wearing your seat belt)?

I'll make another wager and assert that we won't have those kinds of robots sooner than we will have AI self-driving cars, and therefore there really is not much of an incentive to hold-off working on making AI self-driving cars due to the belief that perhaps we'll have driving robots instead.

I have some of my former university research colleagues that keep bugging me about the notion that this whole effort to try and make cars that are AI self-driving cars is a misuse of resources and a misguided effort. They emphasize that if we all put the same energy toward making humanoid robots that had AI based driving skills, we really would not need to change anything at all about cars. Leave cars as they are, they contend. Focus on the "real" problem of making robots. Furthermore, they point out that you can use these humanoid-like robots for a lot more efforts to benefit mankind than a "lousy, single focused AI self-driving car" purpose.

Should we all drop our AI self-driving car efforts and switch over to making robots, ones that could drive our cars, along with those robots being able to clean our houses, run our errands, and possibly even go to work for us and do our jobs at the office? I suppose it is tempting. But, I don't think it is going to dampen anyone's efforts on the AI self-driving car front.

I believe I've sufficiently and "fairly" covered enough about that twist in the plot and will continue unabated about AI self-driving cars that won't be driven by walking and talking robots.

As discussed, it would be handy if we could remove the cars controls from the interior of a car, freeing up the interior space. But, if we remove the car controls, we'd better be sure of what we are doing. This then takes us into the crucial aspects that there are varying levels of AI self-driving cars.

I'd like to introduce the notion that there are varying six levels of AI self-driving cars, ranging from a Level 0 to a Level 5. The topmost level is considered Level 5. A Level 5 self-driving car is a type of self-driving car that is being driven by the AI and there is no human driver involved when the AI is driving the self-driving car.

For the design of Level 5 self-driving cars, some of the auto makers are even removing the gas pedal, brake pedal, and the steering wheel (the human driver controls), since those such auto makers believe that those contraptions are not only unnecessary but furthermore generally unwise to provide. The assertion is that the Level 5 self-driving car is not being driven by a human and nor is there an expectation that a human driver will be present in the self-driving car. It's all on the shoulders of the AI to drive the car.

I'll be revisiting this quite important aspect in a moment about Level 5, so be ready to get into the details, thanks.

A Level 4 self-driving car is somewhat akin to a Level 5, but it is more limited in that it will only be able to drive the car in certain pre-designated circumstances. These circumstances are more formally referred to as Operational Design Domains (ODD).

The ODD's can include such facets as stipulated geographic aspects (similar to doing a geofencing), roadway aspects can be restricted, environmental aspects can be restricted (for example, rain, snow, sleet, etc.), traffic aspects can be restricted (such as only when traffic is open versus crowded), speed aspects can be restricted (a popular restriction is no speeds over 35 miles per hour), temporal restrictions can be specified, and so on.

Any combination of those factors can be established as an ODD. Perhaps an auto maker says that their Level 4 self-driving car will only operate when it is within a 5 mile radius of downtown Los Angeles (a geographic restriction) and only if the traffic is mild (such as not at peak hours of the morning commute and evening commute) and will only go no faster than 35 miles per hour.

Thus, their ODD is this: 5 mile radius + traffic is mild + 35 mph. Some other auto maker might use that same identical ODD, or they might decide to define their own ODD of being a 2 mile radius in any kind of traffic and a max speed allowed of 25 mph.

The auto maker or tech firm that provides a Level 4 self-driving car would presumably let you know beforehand what ODD's or circumstances that entail the AI self-driving car being able to drive the car. As a potential passenger in such a self-driving car, you would certainly want to know under what circumstances it can and cannot do so. Perhaps, for example, the Level 4 self-driving car can handle sunny weather driving, but it is not able to undertake driving in snowy conditions.

Some industry members are worried that the public might get confused by the aspect that various versions of Level 4 self-driving cars will each have their own respective set of ODD's, which are to be designated by the auto maker or tech firm.

In other words, some auto maker we'll call X has come out with a Level 4 self-driving car that can handle self-driving in sunny weather (that's an ODD they could define), and yet it is not able to self-drive in snowy weather (that's another ODD they could define). Meanwhile, a different auto maker, call them auto maker Y, they come out with a Level 4 self-driving car and it can drive also in sunny weather but only on highways and not on narrow city streets (that's another ODD that they could define), and can do self-driving in mild snowy conditions but not severe snowy conditions (that's yet another ODD they could define).

We would now have two different self-driving cars, one by auto maker X and one by auto maker Y, both considered at a Level 4, each of which has defined a proprietary set of ODD's under which their respective self-driving car will operate.

Suppose you are standing at the curb and have called for a ridesharing service to provide you with a self-driving car to get you to the store. One of the Level 4 self-driving cars pulls up at the curb to pick you up. Is it going to be able to drive you around or not? Well, if it is the Level 4 self-driving car provided by auto maker X, you'd better not be headed into snowy conditions. If it is the Level 4 self-driving car provided by auto maker Y, you'd better not be headed into narrow city streets.

It's like the famous line in Forest Gump about the box of chocolates that you never know what you are going to get. The Level 4 is handy because it allows for essentially a more limited version of a Level 5, and presumably allows auto makers and tech firms to progress from perfecting at the Level 4 to then ultimately arrive at a Level 5. But, this also introduces the confusion and potential danger that any Level 4 self-driving car is going to potentially differ from any other Level 4 self-driving car in terms of the nature and range of ODD's or circumstances under which the self-driving car can actually be self-driving.

Imagine a slew of Level 4 self-driving cars, some of them by different auto makers and so the ODD's or allowed driving circumstances differ in that respect. Furthermore, auto maker X comes out with their newest Level 4 self-driving car, an improved version of their older model of their Level 4, and it turns out that the newer model has a new set of ODD's that differ from the prior model.

I realize that some of you will assert that the auto maker X might presumably update the older model via OTA (Over The Air) electronic patches, but we don't know that this will always be the case. In other words, it could be that the older model has various limitations in the hardware that no matter what software updates we might provide, it is still stuck at the prior set of ODD's.

We could end-up then with Level 4 self-driving cars from even the same auto maker that are different in terms of the ODD's or circumstances under which the respective models of their Level 4 self-driving cars can drive. This adds more confusion to the situation. It would be one thing if you somehow knew or were told that the auto maker X's Level 4 self-driving cars can only drive here but not there. Thus, when you saw one of those auto maker X branded Level 4's, you'd know what to expect. It won't necessarily be that way. Ouch!

For self-driving cars at a level 3 and below, there must be a human driver present in the car. The human driver is currently considered the responsible party for the acts of the car. The AI and the human driver are co-sharing the driving task. In spite of this co-sharing, the human is supposed to remain fully immersed into the driving task and be ready at all times to perform the driving task. I've repeatedly warned about the dangers of this co-sharing arrangement and predicted it will produce many untoward results.

Here's the usual steps involved in the AI driving task:
- Sensor data collection and interpretation
- Sensor fusion
- Virtual world model updating
- AI action planning
- Car controls command issuance

We are now ready to focus herein on the driving controls aspects for the various levels of AI self-driving cars.

For the levels 3 and below, there is no question that the driver controls are absolutely needed. The formal SAE definition itself abundantly makes clear that the "fallback" driver must be a human driver. This notion of a fallback refers to the aspect that if the AI decides it can no longer handle the driving task, it can make a request to the present human driver to takeover, plus if the present human driver believes there is a need to take over the driving task from the AI they as the human driver can then choose to do so.

As mentioned earlier, this is a form of co-sharing of the driving task and one that bodes for quite untoward times ahead for the AI self-driving car field. This is also why some auto makers are aiming to skip the Level 3 and instead are aiming solely at the Level 4 and Level 5. In their view, they don't want to get mired in the co-sharing mess that might well be looming upon us.

I trust that you agree with me that the driver controls are needed for Levels 3 and below. This then brings us to the matter of Level 4 and Level 5.

For a Level 4 self-driving car, as earlier mentioned, it will have whatever pre-determined ODD's or circumstances under which it is considered self-driving, as defined by the particular auto maker or tech firm that makes that particular Level 4 self-driving car.

Should a Level 4 self-driving car have human driving controls?

You might at first glance say that it should not. In theory, you should let the AI drive the Level 4 self-driving car. That seems to be the reason to have a Level 4.

But, I ask you this, suppose you are using a Level 4 self-driving car that by the indication of auto maker X cannot be self-driving in snowy conditions. Returning to my earlier example, you have gotten into the Level 4 self-driving car and it is driving you to the store, which is about 15 miles away. During the time that it takes to self-drive you over to the store, snow begins to fall from the sky. Oops!

Now what?

Presumably, according to Level 4, the AI can issue a request for a human driver to intervene and takeover from the AI. If the human driver does not do so, which maybe because there isn't a human driver present in the self-driving car at that moment, or maybe there is a human driver present but they don't want to take over the driving task, or maybe there aren't any humans at all in the self-driving car and the self-driving car is merely on a journey to some other location, etc. If for whatever reason the driving isn't then taken over by a human driver, the Level 4 will perform a fallback operation and try to place the self-driving car into what is called a minimal risk condition (such as parked at the side of the road).

Let's assume that you were in the self-driving car and you are a licensed driver, and the Level 4 self-driving car has asked you to intervene because the snow is falling. Suppose you figure that yes, you'll be happy to go ahead and drive in the light snow, since you grew-up in the Colorado mountains and are quite proficient at driving in the snow.

You inform the AI that you'll gladly take over the driving controls. Wait a moment, I just said that you'll take over the driving controls! If that's the case, and if for Level 4 we allow that they do not necessarily need to have driver controls, how are you going to now take over the driving controls?

Proponents of Level 4 would say that if the auto maker or tech firm had decided to not include driving controls into the Level 4 self-driving car, it's not a big deal because you as the human driver don't need to take over the controls. It is not a necessity that you be ready to take over the controls. It is a nice to have. But, it is not mandatory.

As such, even if you were in that Level 4 self-driving car and there weren't any driver controls, the AI would presumably be aware that you cannot take over the driving, and as such it would perform the fallback to reach a minimal risk condition.

You now are seated in a Level 4 self-driving car that is parked at some edge of a street or highway and you are stranded now because your AI self-driving car has reached the boundaries of what it can do. It cannot in this case handle driving in snowy conditions, and so it has properly brought the self-driving car to a reasoned halt.

How do you feel about those apples?

I dare say you might be irked. If you are a licensed driver and ready to drive, and if you wanted to drive when the Level 4 had reached the end of its ODD set and could no longer self-drive, you are out of luck, buddy. You would sit there and watch the snow falling. You would likely be hoping that the snow will soon stop and thus the AI would re-engage and indicate it can continue the driving journey.

I realize that Level 4 proponents that say the Level 4 should not have driving controls would argue that suppose the situation involved a child in the self-driving car and there wasn't a human driver available. In that case, having the driving controls would be useless anyway. Or, suppose a human driver was available but they were drunk and should not be driving. In that case, perhaps we'd all be thankful there aren't any driving controls in that Level 4 self-driving car.

And so it goes. Each side to this coin has its merits.

You can try to make the case that there should be driving controls in a Level 4, under the belief that when the AI reaches its boundaries of what it can do, there is then the chance that a human driver that is available in that self-driving car, and presumably properly licensed, and sane, and willing to drive can proceed to do so.

Or, you can make the case that the Level 4 self-driving car should not have any driving controls, which will then ensure that you don't get some nutty human driver that opts to take over the controls, plus such controls would not be useful for situations wherein there are only children in the self-driving car or no humans in it, and so on.

There are some AI developers that say that the range of ODD's or circumstances under which a Level 4 will be able to drive the self-driving car will be so wide and deep that it will be a rarity that there would ever need to be a human driver to take over the driving. In essence, if the ODD's cover driving in sunny weather, rainy weather, snowy weather, and at daytime and nighttime, and when on highways and when on city streets, it would imply that this idea of a human driver being needed is rather farfetched. It just won't likely occur, they would assert.

I'd almost buy into that logic except for the aspect that we don't know this to true that Level 4 self-driving cars will have such a robust range of driving capabilities. I would bet that the first rollouts of Level 4 self-driving cars are not going to have a robust range. Instead, it will be a much narrower range. It won't likely be years before we have Level 4 self-driving cars that can meet this claim of being superman-like drivers that can cope with all usual kinds of ODD's or circumstances.

Meanwhile, we are back to my point that once the AI reaches its boundaries on the Level 4 set of ODD's, if we have removed the driving controls then you don't have any chance of having a human driver take over. At least if you had the driving controls, you could presumably then have the chance of a human driver taking over, albeit I realize that the human driver might not be the "best" option per se, depending upon the situation at-hand. It could be that some kind of litmus test be crafted to ascertain whether the human should be able to take over the driving controls, though this is another can-of-worms as to what that litmus test might be and whether it is somehow biased or inappropriate.

Suppose we then all agree that Level 4 self-driving cars should have driving controls in them. Unfortunately, this has adverse consequences and does not solve all of our problems.

By having the driving controls in a Level 4, it takes us back onto the rather untoward turf about the potential of co-sharing the driving task. As already mentioned about Level 3, the co-sharing is dangerous due to the need to convey and communicate between the AI and the human about the driving effort. A wrong hand-off can lead to life-or-death consequences.

You could make the same case about the Level 4 that has driving controls available for a human. On the one hand, you might say that it is entirely different than a Level 3, because the AI engaged in a Level 4 is not going to just routinely handover control to the human driver. The AI would either do so if it determined it had reached the boundary of the ODD's, and notably doesn't actually need a human driver since the AI will otherwise perform the fallback to a minimal risk condition, or the human might opt to disengage the AI and take over the self-driving car.

Here's though where there is a kind of loophole. Suppose the human in the Level 4 self-driving car opts to disengage the AI and takes over the driving, which they can presumably do if there are driving controls. This driver drives for a mile and then decides to re-engage the AI. The AI drivers for a little while, and the human decides to dis-engage it again. The human keeps doing this, opting to sometimes use the AI and sometimes not. This is obviously a kind of separate but co-sharing kind of driving effort, though a rather perhaps disturbing and abrupt kind of driving practice.

If you don't have driving controls in the Level 4 self-driving car, the human cannot try this kind of ping pong match of alternating between AI self-driving and human driving. Instead, the human is relegated to relying entirely upon the AI and will just need to live with the aspect that at times the Level 4 is going to reach its ODD boundaries and come to a reasoned halt.

Some would say that having driving controls in a Level 4 self-driving is going to be like "click bait" (you know, click bait is when on the web they try to trick you into clicking on a web site or link). People that are inside a Level 4 self-driving car are going to be tempted to take over the driving. You know they will. For example, Joe might not like how the AI is driving the self-driving car, such as maybe it is taking him to work, and he is already late getting to work, so Joe takes over the driving to speed along at 90 miles per hour, doing so illegally, which the AI presumably would not do.

By providing the driving controls, you open the can of worms to essentially explicitly proclaim to humans in a Level 4 self-driving car that there are times that they might want to drive the car or maybe even need to drive the car (such as when parked at the side of a road due to the emerging snowy conditions).

Is that the message we want to send?

But, without driving controls, you are then knocking out of the picture the possibility of a human driving, doing so in situations wherein we might all agree that it would have been handy to allow a human to have driven that Level 4 self-driving car.

One such situation involves when a self-driving car gets into an accident of some kind. With conventional cars, you often have a chance at maneuvering a damaged car out of an accident scene by using the steering wheel and perhaps the brakes and accelerator. Without any driving controls in a Level 4, many in law enforcement and the fire departments and auto tow are all concerned as to how a self-driving car is going to be gotten out of the way.

I realize this is a somewhat dizzying matter of trying to decide whether driving controls should or should not be allowed for Level 4 self-driving cars.

I'll bet that you are ready now to consider the Level 5 self-driving cars and whether they should or should not have driving controls. If you are hoping that the Level 5 is a much easier case, well, as mentioned earlier, you might experience some tears rather than joy when I say that it is not an open-and-shut case either.

Let's first establish that the Level 5 is unlike the Level 4 in that the Level 5 must not have any ODD-specific circumstances and therefore is considered "unconditional" in terms self-driving capabilities. This would seem to mean that the Level 5 is always going to be able to drive the self-driving car. But, there is an important small-print caveat that many fail to cite or remember.

As per the SAE standard, a Level 5 self-driving car is supposed to be able to drive the car for all driver-manageable on-road conditions. Notice that there are then two important caveats. One caveat is that the driving circumstances must be driver-manageable. The other caveat is that the driving situations apply only to on-road and not off-road occasions.

We can argue somewhat about what is a driver-manageable circumstance.

I remember that when I was younger, I accidentally slid off a mountain road in my car due to unseen black ice and plowed into a small snow bank. After recovering from the shock of having the car slide without my being able to control it, I tried to back the car out of the snow. I tried and tried but couldn't get it to happen. A highway patrol car came along, fortunately, luckily, and two officers got out of their vehicle to take a look at my predicament. At first, they were going to try and tow me out. One of them bet the other that he could drive it out.

Sure enough, the brazen braggard was able to drive my car out of that snow bank. As the two officers drove off, I'm pretty sure that one was giving the other a few bucks after having lost their own personal bet on whether it was possible to drive out or not. I certainly was happy. I didn't have to wait for a tow truck. I didn't have to pay for a tow truck. I unquestionably didn't feel disappointed in myself that I couldn't drive out my car, since I figured this highway patrol officer had done this a million times and had honed his driving skills for just this kind of moment.

This highlights the difficulty in deciding what is a driver-manageable circumstance. I was sure that my car was stuck in that snow bank and could not be driven out. The other officer was also sure that my car could not be driven out, and he knew a lot more about snow related driving than I did. Yet, it was indeed possible, as we found out when the bragging officer was able to accomplish it.

My snow bank example also encompasses the other caveat of Level 5 self-driving cars, namely that the driver-manageable aspects apply to only on-road situations. Obviously, in my snow bank immersion, I had gone off-road. So, I was in a predicament that covered both being perceived as not driver-manageable and that was unarguably off-road.

Could an AI self-driving car of a Level 5 have been able to drive out of that situation? Per the definition of Level 5, it is not required to be able to do so. Presumably, the AI could be written such that it will try to drive in situations outside the scope of the definition, but there's not a requirement that it do so.

I also don't want you to get misled into thinking that the driver-manageable aspects are only pertinent when an off-road situation arises, which maybe my snow bank example implies.

Here's another personal example of a driving situation and solely illustrates the driver-manageable aspects. I went to a college football game one day and opted to try and park my car in a "fake" parking lot that someone had ingenuously devised to try and make a few extra bucks. It was actually an asphalt basketball court, but an enterprising local had turned it into a parking lot for those desperately searching to find a place to park their car.

The money-making parking lot attendant waved me into an available spot and I thought at the time that I had gotten quite lucky. This impromptu parking lot was just a block away from the stadium.

People parking in the stadium's official parking lot were paying more for their spots and were parking much further away from the actual stadium itself.

After getting settled into my seat for the football game, I got an urgent call that required me to leave the game early. I hurried back over to the impromptu parking lot. When I got there, I stood for a moment in dismay and disgust. The money-maker had packed that asphalt parking lot in such a manner that it maximized his income, but it made things seemingly impossible to get a car out of the pack. I appeared to be completely boxed in.

I suppose that if I had come out once the football game was over, there would be other drivers also wanting to get their cars out of the parking area, and so gradually we could have all figured out a means to undue the parking squeeze-in. In this case, I was the only one that wanted to leave early. The parking lot attendant was nowhere to be seen. He had made his money and hightailed it out of there (probably wasn't even his property to use anyway).

I considered my car to be not driver-manageable in terms of getting it out of that tight knit parking puzzle.

In my mind, the only means to get my car out would be to have a helicopter come overhead and lift it out. As I pondered what to do, a passerby came along and saw me standing there with a look of angst on my face. He stopped to chat and upon sizing up the situation, he told me that he estimated that I could drive the car out – it would require painstaking steps of inching back-and-forth, along with his guiding me by standing outside of my car and waving his hands to signal me to turn the wheel.

I was a bit worried since my car was ding-free and a relatively new car. I thought that maybe he was right that it could be driven out, though I was also pretty sure that I would end-up banging against the other cars in doing so. Turns out that the good news is that with his help, I was able to extricate my car, which took nearly 30 minutes to do. I realize that's a long time to get your car out of a rather miniscule parking lot, but the football game was going to last for another two hours and so it was worthwhile to get my car out when I did (the alternative, waiting for the end of the game, would have been much longer).

I wasn't off-road. I was on a very navigable asphalt lot. My hitch was that I didn't believe the situation was driver-manageable.

Driver-manageable as a measurement is somewhat in the eye of the beholder.

Alright, let's now consider the matter of whether a Level 5 self-driving car should or should not have driver controls.

You might say that a Level 5 self-driving car should not have driver controls. This belief would be based on the notion that there are no ODD's per se for a Level 5 self-driving car, and thus it would seem that the AI will always be able to drive that self-driving car. If that's the case, there's no need for human driver controls.

Also, by removing the driver controls, you are making it clear to any human occupant of the self-driving car that they are fully at the hands of the AI. There is no chance of them trying to take over the controls. Sit back and enjoy the ride, you human.

It also frees up interior space. As mentioned earlier, an auto maker can redesign the interior and make it much more conducive to a wider range of human endeavors, more so than a conventional car or any kind of car that otherwise has driver controls in it.

Rather convincing.

Why would anyone consider putting driver controls into a Level 5 self-driving car?

As mentioned earlier, the AI for a Level 5 only has to be able to handle "driver-manageable" driving situations. Suppose there are situations that appear to be not driver-manageable, and yet could be driven by a human? Likewise, the Level 5 does not need to be able to deal with off-road situations, and yet those off-road occasions might actually be drivable.

In the story of my parking lot spectacle, it could be that the AI might have informed me that my Level 5 self-driving car was stuck in the pack and there was no means to get it out (let's pretend it was a Level 5 self-driving car). That's a fair assessment by the AI and meets the definition of a Level 5 self-driving car in terms of driving capabilities. I would have been stuck with a multi-ton paperweight and would have had to wait until the football game ended and people came to move their cars.

In the story of my snowbank debacle, it could be that the AI might have informed me that my Level 5 self-driving car was stuck in the snow and it was considered off-road. That's a fair assessment by the AI and meets the definition of a Level 5 self-driving car in terms of driving capabilities. I would have likely had to wait for a tow truck to yank my car out of the snow or perhaps get the officers to try and tow me out of the snowbank with their patrol car.

Notice though that in both of those stories, in the end, a human driver was actually able to drive the car successfully. If the car had been an AI self-driving car of a Level 5, the topmost level of the scale, and if there had not been any driver controls available, no human driver would have been able to drive the stuck car out of its mired position.

I am sure that some pundits will be yowling that my examples are obscure, and they perhaps overplay the aspects of what constitutes driver-manageability and also what constitutes off-road. I'm not so sure their claims in that regard are quite so valid.

There are some too that say that a Level 5 is not precluded from driving while off-road, meaning that the definition only says that it does not have to be able to do so. Furthermore, we can presumably argue until the cows come home as to what a driver-manageable situation consists of. Pundits would say that if the average adult driver can drive it, that's considered a driver-manageable situation.

Also, pundits would say that with the use of OTA, it would be feasible that your Level 5 self-driving car might be able to get a downloaded system patch or special update that would allow it to drive like the highway patrol officer. In other words, in my case as a normal adult driver and not being able to drive out of the snowbank, presumably a bit of added program code could be downloaded into my Level 5 self-driving car and the add-on would make my Level 5 self-driving car as proficient as the highway patrol officer.

In that case, one could argue that AI self-driving cars will one day be better, likely much better than the average adult driver and be able to negotiate all sorts of seemingly non-manageable driving situations due to having superior AI driving skills.

Yes, it could be that ultimately and eventually we'll have AI self-driving cars that are masters at the driving task. I'd dare say though that we are not going to get there right away. Instead, there will be purported Level 5 self-driving cars that are not particularly as robust a driver as the average adult driver and therefore we'll be witnessing those Level 5 self-driving cars get into situations that are human driver manageable but not strictly-speaking AI driver-manageable.

We should also add to this argument that just as with Level 4 self-driving cars, if a Level 5 self-driving car gets into an accident, it is unclear whether the Level 5 is going to be able to extricate itself from the accident. Is that considered a driver-manageable task?

Most would say that once an AI self-driving car has gotten into an accident, it is considered then no longer bound by the rules of being driver-manageable. If that's the case, how will first responders deal with the Level 5 self-driving car? Might the then paperweight rendered Level 5 self-driving car be a hazard to others and the traffic, even if it perhaps is still drivable, but just not so by the AI?

I'll also chuck into this debate the aspect of whether the AI self-driving car at a Level 5 or even a Level 4 might have some kind of bug in it that makes it undrivable by the AI. You might say that the Level 4 or Level 5 is supposed to be able to self-diagnose and realize it is in a buggy state and then aim toward the fallback and a minimum risk condition.

This ability though to even realize it needs to do a fallback could be part of the bug. Likewise, the bug could be in the system portion that is trying to get the self-driving car into a minimum risk condition. Under such circumstances, it might be handy to have driver controls, albeit again the other earlier issues about whether there is a proper human driver available to make use of them.

Have I now convinced you that the idea of not having driver controls in a Level 5 self-driving car is at least something worthy of re-considering?

I ask you this aspect because there are some that are absolutely convinced that a Level 5 self-driving car should not have driver controls. When you ask them why they believe this, they usually don't really know. It just seems that they've bought into a futuristic theme of not having driver controls. For those of you that have more carefully pondered the pro's and con's of having driver controls, I applaud you for trying to figure this out with some logic and aplomb.

Let's shift the discussion now towards other relevant tangents associated with this question about driver controls.

Some might say that if you were to allow for a remote operator that could drive an AI self-driving car, you would not need driving controls inside the self-driving car and yet could still allow for a human to drive the self-driving car when needed. This then seems to solve our dilemma. You leave out the driving controls that are normally within the self-driving car, which frees up the interior space and effectively proclaims to the human occupants that they are not going to be driving the car, ever.

Sorry to say that I don't buy into the remote operator approach. There are too many downsides. In what circumstances would the remote operator be able to take over the driving control? Can the remote operator really properly be able to "see and sense" the surroundings of the self-driving car to be able to suitably and safely drive the self-driving car? If the self-driving car is otherwise already damaged, would the remote operator have full sensory capability at-hand? Suppose the self-driving car is out of range of electronic communications and the remote operator cannot connect with it? And so on.

I've debunked the remote operator as a savior concept and have doubts that it will become a big thing, which I realize that some pundits are already claiming it will be and are hiring like mad to staff-up for thousands upon thousands of remote driver operations for self-driving cars. We'll see how that pans out.

Who knows, maybe your next job will be as a remote driver operator for a rapidly expanding AI self-driving car company. Good luck on that.

Okay, let's consider some other alternatives about the driving controls conundrum.

We have so far assumed that driver controls consist of a steering wheel, a brake pedal, and an accelerator pedal. Those are the customary kinds of driver controls.

Maybe we should reinvent the driver controls.

Let's imagine that the AI self-driving car of a Level 5 has parked in that basketball court asphalt parking lot. It did so when it was easy to park there, since there weren't other cars blocking it or making it hard to do so. You come out to your Level 5 self-driving car and plead with it to try and extricate itself from the parking lot. It tells you that it cannot do so and has gotten boxed in. The AI explains to you that the self-driving car is now in a non-manageable driving situation.

Remember when I told the story earlier and I had mentioned that a passerby helped me to get out of the parking lot, doing so by giving me a series of inch at a time instructions?

We can use that same concept with the Level 5 self-driving car.

Perhaps the AI ought to let a human provide driving instructions to it, allowing a human-assisted or human-guided driving act to occur when the AI has gotten stumped on a driving task. You might for example stand outside of your Level 5 self-driving car and tell it to back-up an inch, then tell it to stop, then tell it to move the wheel to the right, then stop, and so on. You could converse with the AI to help provide an indication of how to drive the car.

In that manner, even if the AI had reached the end of its driver-manageable capability, it could be augmented by a human. Is the human now driving the car? Not really. The AI is still driving the car. In that sense, we might say that the AI is still "responsible" for the driving act. I mention this point because once we've opened the pandora's box of having the human instruct the AI on how to drive, we are confronted with all sorts of other potential problems. Suppose that the human tells the AI to drive the self-driving car into a wall or over a cliff?

Anyway, the point being that you don't necessarily need to have physical driving controls inside the self-driving car and could potentially instead allow a human to instruct the AI on how to drive the self-driving car. There are numerous pro's and con's to this approach and I'd like to make clear that it is not some kind of slam dunk solution. I've written and spoken extensively about the trade-offs and won't repeat those other facets herein.

One counter-argument about this notion of a human conversing with the AI to have the AI then drive at the directed instructions of the human involves the possibility that the AI itself is out-to-lunch and therefore unable to obey your commands. Suppose the AI is buggy or has gotten busted in an accident, and you want to try and drive the car but the AI is so messed-up that it won't listen to you or gets confused by your commands – it's a Catch-22 because the AI is not going to be responsive to your commands and yet its the only means allowed for you to drive the car.

This is why some argue that physical driving controls are a better choice (though, one can of course argue that even physical driving controls can breakdown too).

Since we are on the topic of being able to communicate with the AI self-driving car, and perhaps do so in a human-guided fashion, I'll bring up another option that's a bit farfetched for some. Rather than you speaking to the AI of the self-driving car, and rather than maybe typing messages into a console or texting them to the AI, suppose instead that you could "think" to the AI? By this I mean that the AI could somehow read your thoughts. Remember I forewarned you this was a stretch.

In any case, there are increasing advances in being able to read various brainwaves. It is believed that someday we'll be able to communicate with automata and other humans via thinking alone. I've written and spoken about this matter and won't repeat all the pro's and con's herein but suffice to say that it is certainly an interesting concept and one that maybe one day we'll live to see happen. This futuristic approach is commonly referred to as brainjacking.

Another idea is that maybe the driving controls would be physically present in the Level 5 self-driving car but be foldable or hidden from use when so desired. Imagine that an auto maker made a steering wheel that folds into the dashboard and is no longer readily visible or usable. When you need to use the steering wheel, you can open the dashboard panel that houses the steering wheel and pop the steering wheel into place. The same would be the case with the brake pedal and the accelerator pedal.

Here's the way this would presumably work. You are waiting for a Level 5 self-driving car to come pick you up to take you to the store. It arrives. There does not appear to be any visible driver controls. You get into the Level 5 self-driving car. Away it goes. It parks at the store and awaits your finishing your shopping trip. When you come out with all of those bags of groceries, you notice that your shiny new Level 5 self-driving car has gotten boxed in (similar to my story about parking for the football game).

The Level 5 self-driving car indicates to you that it cannot do anything because the situation is not driver-manageable. You opt to get into the Level 5 self-driving car and tell it that you are going to take over the driving. You then foldout the steering wheel and pedals and proceed to extricate the self-driving car from the parking spot, doing so entirely on your own and without any assistance by the AI. Once you've done so, you fold back in the driving controls and tell the AI to take over the driving task henceforth to get you home.

I'll point out that this is again not a slam dunk kind of solution. Under what circumstances should these hidden driving controls be allowed to be used? Suppose a child is alone in the self-driving car and opts to foldout the driving controls and take over the driving? I think you can see that this has a number of tradeoffs as a solution.

I'd like to mention one interesting aside about these situations wherein a human might be guiding the AI of a self-driving car. You could say that such moments are teachable moments. In essence, maybe we should use those situations to have the AI get better at driving the self-driving car. If that was the case, the number of instances in which the AI got stumped would presumably gradually diminish, since its repertoire of driving skills would be increasing.

How else might we deal with driver controls?

It could be that an auto maker might decide to offer two different models of their Level 5 self-driving car. One version has driver controls, the other version does not. This then leaves the choice up to the consumer or buyer of the self-driving car. If the purchaser thinks that having the driving controls is an advantage, they will buy that version. If the purchaser believes that having driving controls is not needed, they will buy that version.

We might therefore let the marketplace decide. This though has its own tradeoffs. Would the buyer even understand the significance of getting or not getting the driver controls? Would the aspect that some version of the same Level 5 car has driving controls while others do not be overly confounding and confusing for all?

For the auto maker, it would also be a pain-in-the-neck. The cost to have two different models will certainly heighten the overall costs of their Level 5 self-driving cars. The manufacturing and assembly are bound to be different. The AI would likely also need to differ in terms of the version for the driving controls versus the one that does not have the driving controls. Probably a bit of a nightmare.

Of course, auto makers do have variants of their conventional cars, ones that offer one set of features versus a different set, so this idea of omitting or including driver controls is not totally farfetched.

Speaking of features, you should consider the Level of the self-driving car as essentially being a kind of feature of the car. Here's why.

Keep in mind that the Level 4 and Level 5 standard definitions apply to the self-driving car when its AI system is engaged, meaning that presumably you don't necessarily need to engage the AI. So, if I tell you that a particular self-driving car is a Level 5, I mean that it is able to have the AI do the self-driving at the Level 5 level, but only when the AI self-driving system is engaged.

This brings up a crucial question for the auto maker, namely, should they or should they not allow for the AI on-board system to be able to be disengaged at the whim of a human?

An auto maker might decide that the only means of the Level 4 or Level 5 self-driving car being usable at all will be when the AI driving is engaged and otherwise their Level 4 or Level 5 self-driving car is not going to budge an inch. Thus, in that sense, the AI system is considered always engaged and never able to be disengaged, with respect to the driving of the car.

A different auto maker might decide that their Level 4 or Level 5 self-driving car is able to be used by a human driver whenever the human wants, in which case the AI on-board system is to be disengaged first, prior to the human driver taking on the driving controls. When the human decides they don't want to drive the self-drivable car at some point, the human then engages the AI system.

This act of engagement and disengagement might involve pressing a big red button that's inside the car (well, it could be any kind of button or similar kind of on/off switch), or it might involve sending a signal to the self-driving car via your smartphone, or you might verbally tell the self-driving car that you want the AI engaged or disengaged.

Some believe that a Level 4 or Level 5 self-driving car should always be driven by the AI and never be allowed to be driven by a human, therefore the AI is considered always engaged and cannot be disengaged no matter what. Others suggest that if a person wants to have and use a Level 4 or Level 5 self-driving car and do so by driving it themselves, well, they ought to be able to do so. The human is able to decide whether they themselves want to drive the car or let the AI do the driving.

I think you can see that this is another related point of contention in the debate over driver controls.

If you've been with me throughout this whole journey of debating the inclusion or exclusion of driver controls, I applaud your tenacity. As I mentioned at the start of this discussion, it's a very popular question and one that unfortunately does not have a twitter-sized answer per se, other than to say "it depends."

For now, I'll leave you with one final thought on this matter.

Will people be willing to give up the use of driving controls?

I ask this because we might devise all sorts of clever ways to avoid having to provide driving controls, and yet in the end it might be that people insist on having them. I realize this idea of letting people have driving controls is somewhat antithetical to the Utopian world of a future that has presumably no car accidents because we have eliminated the human driver from the equation of driving.

There are some that keep insisting we will have zero fatalities due to the advent of AI self-driving cars.

For the Level 3 and below, there are still going to be car crashes and deaths, since the car driving is co-shared between humans and the AI. Even for the Level 4 and Level 5, there will still be car crashes and deaths. How can that be?

Keep in mind that there are avoidable car crashes and unavoidable car crashes. An example of an unavoidable car crash is when a pedestrian suddenly and unexpectedly steps off the curb into the path of a car that is going say 45 mile per hour and does so with only a split second of reaction time. No human driver and no AI driver is going to be able to avoid such an unavoidable car crash. Hopefully, we'll have a lot less of those unavoidable instances, but in any case, it isn't going to drop to zero.

Anyway, let's go with the notion that if we did have an all and only Level 5 world of AI self-driving cars, and denied humans from driving, it would hopefully eliminate the drunk driving and other such human foibles of driving that leads to deaths and injuries, and we would dramatically decrease the number of annual deaths due to driving incidents.

I'll repeat my question, namely will people be willing to live in such a world without any human allowed driving controls, or might they insist that you'll remove their driving controls over their dead bodies (a play on the "you'll remove my gun over..."). Would the government and regulators be able to mandate that driving controls are no longer permitted? Maybe the public would accept this notion in return for the decreased number of deaths. I'm betting though that a segment of society will still believe in the "right" to be able to drive a car (yes, I realize it is considered a privilege rather than a right, but I'm saying members of the public-at-large might consider it to be a right).

Perhaps society accedes to those that want to have driving controls, but they then can only use those driving controls in certain settings. If you are on private land, maybe you can use them there, but not when on public roads. If you attempt to use them when they are not legally permitted, either the AI stops you from doing so, or if you get caught you get a hefty ticket and maybe jail time. That kind of thing.

From an AI perspective, much of this debate about the driving controls on AI self-driving cars deals with a much larger macroscopic question about AI systems overall. Allow me to explain.

As more AI systems get devised and rolled out, we need to be aware of the AI boundaries problem. Simply put, when the AI has reached the end of its rope and can no longer perform whatever task is involved, what happens then? Do we hand the situation over to humans? In what way is this best done? Will the humans know what they are doing?

Along those lines, there are some that are worried that with a mix of conventional cars and AI self-driving cars, and if we eliminate the driving controls on some of the AI self-driving cars, it could imply that gradually people that are actually able to drive will have their driving skills decay because they are becoming reliant on the AI to do the driving. If someone is used to being chauffeured most of the time, and then they are suddenly asked to drive a car themselves, the odds are that their driving skills are lessened. This could also mean that they might be more apt to get into car accidents and car crashes.

You can say the same about most other tasks that an AI system might be devised to undertake. Suppose we put an AI robotic arm in a manufacturing plant and humans no longer do what the robotic arm does. If for some reason you need a human to do that task, will humans be available that know how to do so? Will they still be proficient? Might they and others get injured due to the lack of proficiency?

Some are worried that we are heading toward a world of AI that dominates what we do, and we as humans will gradually lose our own skills at doing things. When the AI falters or falls apart, humans will be stumped as to how to do those tasks. We are going to by default allow AI to let humans become deskilled.

In the case of the car driving task, it is important because the AI is not yet able to truly do whatever a human driver can do. We are a long way from that happening. In the meantime, we'll be using AI that is relatively brittle and narrow in what it can do as a driver. Will we meanwhile become deskilled as drivers, and yet think that we can jump behind the driver's wheel and drive if we wish to do so? It could be that the deskilling on a mass scale would lead to havoc as those drivers that were once seasoned have reverted to being novices.

Don't want to scare anyone about that AI boundaries issue, and I'm not saying the sky is falling, but it is something that as AI developers we should be considering. I'm a proponent of the belief that AI developers should take responsibility for what they do and what they are making, and I ascribe to various professional association codes of conduct that argue for that avowal.

Guess what? You are now officially indoctrinated into the ongoing debate about car driving controls.

Congratulations! You made it through the quagmire and can now get involved directly into the at-times acrimonious dialogue. If you were looking for a quick answer about whether or not driving controls should be included or should be excluded from the topmost levels of self-driving cars, my answer is yes.

Wait, was Lance saying that yes they should be included or yes they should be excluded.

Ask the AI what it thinks.

CHAPTER 3

BUG BOUNTY AND
AI SELF-DRIVING CARS

CHAPTER 3

BUG BOUNTY AND
AI SELF-DRIVING CARS

Bounty hunter needed to find a copper pot that went missing from a small shop. Reward for recovery of the copper pot will be 65 bronze coins. Thusly said a message during the Roman Empire in the city of Pompeii. We don't today know if any bounty hunter found the copper pot and claimed the bronze coins, but we do know that bounty hunting dates back to at least the times of the Romans.

In more modern times, you might be aware that in the 1980s there were some notable bounties offered to find bugs in off-the-shelf software packages and then in the 1990's Netscape notably offered a bounty for finding bugs in their web browser. Google and Facebook had each opted toward bounty hunting for bugs starting in the 2010 and 2013 years, respectively, and in 2016 even the U.S. Department of Defense (DoD) got into the act by having a "Hack the Pentagon" bounty effort (note that the publicly focused bounty was for bugs found in various DoD related web sties and not in defense mission critical systems).

According to statistics published by the entity HackerOne, the monies paid out in 2017 toward bug bounty discoveries totaled nearly $12 million dollars and for 2018 it is sizing up to be more than $30 million dollars. For bugs that are considered critical issues, the usual bounty is around $2,000. Bounties though are decided by the eye of the beholder in the sense that whomever is offering the bounty might go lower or higher and in some cases there have been bounties supposedly in the $250,000 range.

Some are puzzled that any firm would want to offer a bounty to find bugs in their software. On the surface, this seems like "you are asking for it" kind of a strategy. If you let the world know that you welcome those that might try to find holes in your software, it seems tantamount to telling burglars to go ahead and try to break into your house. Even if you already believe that you've got a pretty good burglar alarm system and that no one should be able to get into your secured home, imagine asking and indeed pleading with burglars to all descend upon your place of residence and see if they can crack into it. Oh, the troubles we weave for ourselves.

Those that favor bounty hunting for software bugs are prone to saying that it makes sense to offer such programs. Rather than trying to pretend that there aren't any holes in your system, why not encourage holes to be found, doing so in a "controlled" manner? In contrast, without such a bounty effort, you could just hope and pray that by random chance no one will find a hole, but if instead you are offering a bounty and telling those that find a hole that they will be rewarded, it offers a chance to then shore-up the hole on your own and then prevent others from secretly finding it at some later point in time.

Well-known firms such as Starbucks, GitHub, AirBnB, America Express, Goldman Sachs, and others have opted to use the bounty hunting approach. Generally, a firm wishing to do so will put in place a Vulnerability Disclosure Policy (VDP). The VDP indicates how the bugs are to be found and reported to the firm, along with how the reward or bounty will be provided to the hunter. Usually, the VDP will require that the hunter end-up signing a Non-Disclosure Agreement (NDA) such that they won't reveal to others what they found.

The notion of using an NDA with the bounty hunters has some controversy. Though it perhaps makes sense to the company offering the bounty to want to keep mum the exposures found, it also is said to stifle overall awareness about such bugs. Presumably, if software bugs are allowed to be talked about, it would potentially aid the safety of other systems at other firms that would then shore-up their exposures. There are some bounty hunters that won't sign a NDA, partially due to the public desire and partially due to trying to keep their own identity hidden. Keep in mind too that the NDA aspect doesn't arise usually until after the hunter claims they have found a bug, rather than requiring it beforehand.

Some VDP's stipulate that the NDA is only for a limited time period, allowing the firm to first find a solution to the apparent hole and then afterward to allow for wider disclosure about it. Once the hole has been plugged, the firm then allows a loosening of the NDA so that the rest of the world can know about the bug. The typical time-to-resolution for bounty hunted bugs is usually around 15-20 days when a firm wants to plug it right away, while in other cases it might stretch out to 60-80 days. In terms of paying the bounty hunter, the so-called time-to-pay, after the hole has been verified as actually existing, the bounty payments tend to be within about 15-20 days for the smaller instances and around 50-60 days for the larger instances.

Who are these bounty hunters? They are often referred to as white hat hackers. A white hat hacker is the phrase used for "hackers" that are trying to do some kind of good. We normally think of hackers as cybersecurity thieves that hack their way into systems to steal and plunder. Those are usually considered black hat hackers. Consider that hacking is akin to the days of the Old West, wherein the good gun slingers wore white hats and the evil ones wore black hats (well, that's what TV and movies suggest).

For anyone that knows much about hacking, such as trying to break into a system, it is somewhat frustrating that the mass media will often confuse true hacking from marginal hacking. If someone uses a social engineering technique to get your password, perhaps calling you on the phone and claiming to be with tech support and asking you for your password, few "genuine" hackers would consider that to be a form of hacking. The culprit merely tricked someone into giving up their password.

If instead the culprit had used some kind of password cracking program that they had written, or if they found some exploitable bug in the password entry program, it would give them more credence as a hacker. It used to be that most of the true hacking was being done by hard-core programmers that knew the inner sanctum aspects of various operating systems and other software. Lately, just about anyone can either use social engineering or can purchase via the dark web various cracking programs that need only to be run. These less bona fide hackers often have very little computer skills and sometimes don't even know how to write a line of code.

This brings us to the topic of what kinds of software bugs the bounty efforts are looking for. Generally, the bounty program excludes things like social engineering. It's more about having identified an actual bug in the system. The bounty hunter normally has to be relatively clever and try all sorts of potential exploits to find a hole. It can be a laborious process. There is no guarantee that the bounty hunter will find any holes. This doesn't mean that there aren't any holes, it just means that the bounty hunter couldn't find them.

A firm might feel better about its software if dozens or perhaps hundreds or thousands of bounty hunters have tried to find software bugs and have not been able to do so. Again, this is not any kind of proof that no such bugs exist. But, if these multitude of efforts do not bring forth a bug, it would seem to suggest that they are either not there or perhaps very hard to find. This might imply that someone else of a dishonorable nature that comes along later on, not having anything to do with the bounty effort, will be unlikely to also find any bugs.

Suppose a bounty hunter finds a bug but decides not to tell the firm? That's the classic conundrum.

If the firm provides a "safe harbor" protection via their VDP, meaning that they will not try to go after the bounty hunter for finding a bug, and if the firm offers enough of a monetary incentive, the bounty hunter is hopefully swayed toward reporting the bug to the firm.

On the other hand, the bounty hunter might be both a white hat and a black hat kind of hacker, such that if the bug is an exposure that could be exploited to steal or plunder, the value of the bounty might be insufficient and so the hunter keeps the bug under wraps.

The bounty hunter though that keeps secret about the bug in hopes of later utilizing it for some nefarious act will also then become potentially exposed to adverse legal repercussions, either by the firm suing them if they act upon the bug or possibly even have criminal charges aimed at them. And, the bounty hunter has to wonder whether perhaps some other bounty hunter might find the bug, in which case, the other bounty hunter will potentially claim the prize over them.

Often, for bounty efforts, more than one bounty hunter finds the same bug. The firm that is undertaking the bounty effort needs to figure out which of the bug reports are duplicative. They also need to figure out which bounty hunter should get the credit for having found the bug. In many cases, the bounty hunters use some kind of reporting system setup by the firm to indicate the bugs being found, and as a result the logging keeps track of which bounty hunter first reported the bug.

I've worked with companies that thought doing a bug bounty effort would be a "fun" and publicity worthy activity. I pointed out to them that beyond the aspects aforementioned about the possible dangers of doing such an effort, it also often produces lots of false reports. In essence, there are bounty hunters that are desperate to try and win some of the bounty and so they will log all sorts of wild things that are not bugs at all.

In the days of the Old West, suppose you offered a reward for the capture of Billy the Kid (a famous outlaw). If you did so and did not include a picture of what Billy looked like, imagine the number of bounty hunters that might drag into the sheriff's office someone that they hoped or thought was Billy the Kid. You might get inundated with false Billy's. This is bad since you'd need to presumably look at each one, asking probing questions, and try to ascertain whether the person was really Billy or not.

The same is the case for scrutinizing the bounty hunter submissions. There will be a lot of "noise" in the reported bugs, in the sense that many of the claimed bugs don't exist, and the bounty hunter just thought they found one.

Unfortunately, being able to determine which of the reported bugs are valid and which ones are not will take a lot of laborious effort by your highly skilled software engineers. It means that they will be taken away from whatever else that they should be doing. I mention this because there is a substantive cost involved in assessing the bugs, and many firms don't account for that cost when they decide to run one of these bounty efforts. They naively seem to think that only bona fide bugs will be reported. Not so.

If you are pondering what kind of bugs might be found, you can take a look at the Common Vulnerability Scoring System (CVSS) to see how bugs are labeled as either low, medium, high, or critical, along with seeing examples of such bugs. One example that is easy to describe is labeled as CVE-2009-0658 and involves the Adobe Acrobat buffer overflow vulnerability (which has since been fixed).

Essentially, if you tried to open a PDF document that contained a malformed picture (one likely purposely malformed), it would cause an overflow in the Adobe software buffer and allow a remote attacker to be able to then executive code on your system. This would be especially attractive to the interloper if you happened to have system privileges on your machine, and thus by opening the devious PDF in your Adobe Reader you would have opened up pandoras box. Based on a combination of metrics including the attack complexity, user interaction required, and so on, it earned a CVSS v2 base score of 9.3.

In some cases, the firm doing the bounty program will make it open to the public. Anyone that wants to have at it, please do so. These are usually time-bounded. The firm will declare that the bounty program starts say a month from now and will last for 60 days. This helps to then spark interest and get those bounty hunters looking. There are also time un-bounded bounty programs, wherein a firm will at any time welcome a bounty hunter offering a proposed found bug.

During the days of the Old West, this kind of open call would often bring forth vigilantes and bounty hunters that had no idea what they were doing. It was a free for all. As such, some of the software bug bounty programs are at times public but still restricted in some fashion. For example, you might need to officially register with the bounty effort and provided some kind of evidence of your credentials.

There are also private oriented bounty efforts. In the private instances, the firm will tend to seek out specific known white hat hackers and arrange for them to get access to the software that is going to be put through the wringer. This hopefully reduces too the chances of a black hat hacker getting involved.

There is an ongoing debate in leadership circles about whether it is better to use a bounty approach or to instead hire a bug-finding firm to do the work instead. There are plentiful number of firms that will do security threat analyses and do the same kind of work that bounty hunters would do. You can establish the hourly rate or a set fixed price for them to assess your systems and try to find bugs. They can then work hand-in-hand with your software team and it is all done as a rather confidential matter.

There are those that argue that you cannot possibly pay the same token that you would pay when doing bounty hunting. In other words, there might be hundreds of bounty hunters spending gobs and gobs of hours trying to find bugs. One of the bounty hunters finds a bona fide bug and you pay that person say $1,500. If you had been paying specialists to search for bugs, it might have cost you $15,000 or maybe $150,000 to have found that same bug. Thus, in theory, the bounty approach is a cheaper way to find bugs (maybe!).

Some would even argue that your own internal software team should be doing the bounty hunting. I've had some lengthy discussions about whether to offer a "bonus" to any member of the team that finds a bug, which can unfortunately also produce counter-productive behavior. In one firm, the team members were planting bugs to be able to get bonuses when they found the bugs. This is not in the spirit of such an effort and there are ways to try and avoid getting into such an awkward and untoward predicament.

One argument against using your own team to find bugs is that they are too familiar with the software to potentially find the bugs. They wrote the software and so might make all sorts of assumptions that would blind them to finding bugs. By using outsiders, the outsiders are trying all kinds of wild tricks to find bugs. They don't know where the bugs are. They use their outsider lack of awareness to try all avenues, and don't assume that you must have done various testing and safeguards. The counter-argument is that you should simply divide your own developers into a blue team, red team, and sometimes a purple team, and thus gain a somewhat similar sense of outsider assessments.

What does this have to do with AI self-driving cars?

At the Cybernetic AI Self-Driving Car Institute, we are developing AI software for self-driving cars. Besides our own efforts to find and eliminate any potential bugs, we also are able to aid other tech firms and auto makers by being private "bounty hunters" when requested, focusing on specifically AI self-driving car systems.

A macroscopic question though is whether or not the auto makers and tech firms should use bounty hunter efforts or not?

Similar to my earlier points, you might at first say that of course the auto makers and tech firms that are making AI self-driving cars should not undertake public oriented bounty hunter programs. Why would they allow hackers to try and find bugs in AI self-driving car systems? Isn't this tantamount to having your home examined closely by burglars? In fact, it's scarier than that. It's like having an entire neighborhood of homes closely examined by burglars, and they might not just be interested in your jewels and money but maybe be a threat to your personal safety too.

When you consider that AI self-driving cars are life-or-death systems, meaning that an AI self-driving car can go careening off the road and kill the human occupants or humans nearby, it would seem like the last thing you would want to do is invite potential black hat hackers to find holes.

The counter-argument is that if the auto makers or tech firms don't do a bounty type program, will they end-up putting on the roads an AI self-driving car that has unknown bugs, for which the black hat hackers will ultimately find the holes anyway. And, once those holes are found, the dastardly results if exploited could be life-and-death for those using the AI self-driving cars and those nearby them.

I'd like to clarify and introduce the notion that there are varying levels of AI self-driving cars. The topmost level is considered Level 5. A Level 5 self-driving car is one that is being driven by the AI and there is no human driver involved. For the design of Level 5 self-driving cars, the auto makers are even removing the gas pedal, brake pedal, and steering wheel, since those are contraptions used by human drivers. The Level 5 self-driving car is not being driven by a human and nor is there an expectation that a human driver will be present in the self-driving car. It's all on the shoulders of the AI to drive the car.

For self-driving cars less than a Level 5, there must be a human driver present in the car. The human driver is currently considered the responsible party for the acts of the car. The AI and the human driver are co-sharing the driving task. In spite of this co-sharing, the human is supposed to remain fully immersed into the driving task and be ready at all times to perform the driving task. I've repeatedly warned about the dangers of this co-sharing arrangement and predicted it will produce many untoward results.

Let's focus herein on the true Level 5 self-driving car. Much of the comments apply to the less than Level 5 self-driving cars too, but the fully autonomous AI self-driving car will receive the most attention in this discussion.

Here's the usual steps involved in the AI driving task:
- Sensor data collection and interpretation
- Sensor fusion
- Virtual world model updating
- AI action planning
- Car controls command issuance

Another key aspect of AI self-driving cars is that they will be driving on our roadways in the midst of human driven cars too. There are some pundits of AI self-driving cars that continually refer to a utopian world in which there are only AI self-driving cars on the public roads. Currently there are about 250+ million conventional cars in the United States alone, and those cars are not going to magically disappear or become true Level 5 AI self-driving cars overnight.

Indeed, the use of human driven cars will last for many years, likely many decades, and the advent of AI self-driving cars will occur while there are still human driven cars on the roads. This is a crucial point since this means that the AI of self-driving cars needs to be able to contend with not just other AI self-driving cars, but also contend with human driven cars. It is easy to envision a simplistic and rather unrealistic world in which all AI self-driving cars are politely interacting with each other and being civil about roadway interactions. That's not what is going to be happening for the foreseeable future. AI self-driving cars and human driven cars will need to be able to cope with each other. Period.

Some say that it would be dubious and actually dangerous for the auto makers and tech firms to consider doing a public oriented bounty program for finding bugs in AI self-driving cars. If those entities want to do a private oriented bounty program, involving carefully selected white hat hackers, it would seem more reasonable given the nature of the life-and-death systems involved.

It becomes on the heads of the auto maker or tech firm then whether using a private bounty program is best, or whether to instead hire a firm to do the equivalent, or whether to try some kind of internal bounty effort. The presumption is that the auto maker or tech firm needs to decide what will most likely reduce the chances of bugs existing in the AI self-driving car systems. In fact, the auto maker or tech firm might try all of those avenues, doing so under the notion that given the importance of such systems and their critical nature, the more the merrier in terms of finding bugs.

There are some that believe that the auto makers and tech firms might not take seriously the need to find bugs and thus the assertion is made that regulations should be adopted accordingly. Perhaps the auto makers and tech firms should be forced by regulatory laws to undertake some kind of bounty efforts to find and eliminate bugs. This is open to debate and for some it is a bit of an overreach on the auto makers and tech firms. It is likely though that if AI self-driving cars appear to be exhibiting bugs once they are on our streets, the odds are that regulatory oversight will begin to appear.

One view is that there's no need to do a large-scale casting call for finding bugs. Instead, the AI self-driving cars themselves will be able to presumably report when they have a bug and let the auto maker or tech firm know via Over The Air (OTA) processing. The OTA is a feature for most AI self-driving cars that allows the auto maker or tech firm to collect data from an AI self-driving car, via electronic communication such as over the Internet, and then also be able to push data and programs into the AI self-driving car.

It is assumed that the auto makers and tech firms will dutifully and rapidly send out updates via OTA to their AI self-driving cars, shoring up any bugs that are found. Though this is supposed to be the case, there will still be a time delay between when the bugs are discovered and then a bug patch or update is prepared for use. There will be another time delay between when those patches get pushed out and when the AI self-driving cars involved are able to download and install the patch.

I mention this time elapsed periods because some pundits seem to suggest that if a bug is found on a Monday morning at 8 a.m., by 8:01 a.m. the bug will have been fixed and the fix sent to the AI self-driving car. Not hardly. The auto maker or tech firm will need to first determine whether the bug is really a bug, and if so what is causing it. They will need to find a means to plug or overcome the bug. They will need to test this plug and make sure it doesn't adversely harm something else in the system. Etc.

Even once the patch is ready, sending it to the AI self-driving cars will take time. Plus, most of the AI self-driving cars are only able to do updates via the OTA when the AI self-driving car is not in motion and in essence parked and not otherwise being active. If you are using an AI self-driving car for a ridesharing service, the odds are that you'll be running it as much as you can, nearly 24x7. Thus, trying to get the OTA patch will not be as instantaneous as it might seem.

We also need to consider the severity of the bug. If the bug is so severe that it causes the AI self-driving car to lose control of the car, such as if the AI freezes up, you are looking at the potential of an AI self-driving car that rams into a wall, or slams into another driver, or rolls over and off-the-road. The point being that you cannot think of this as finding bugs in perhaps a word processing package or a spreadsheet package. These are bugs in a real-time system and one that holds in the balance the lives of humans.

For those of you that pay attention to the automotive field, you likely already know that General Motors (GM) was the first auto maker to formally put in place a VDP, doing so in 2016. For their public bounty efforts, the focus has tended to be the infotainment systems on-board their cars or other supply chain related systems and aspects.

Overall, it has been reported that GM from 2016 to the present has been able to resolve over 700 vulnerabilities and done so in coordination with over 500 bounty hunters and hackers. Within the GM moniker, this effort includes Buick, Cadillac, Chevrolet, and GMC. Currently, an estimated seven of the Top 50 auto makers have some kind of bounty program.

This is overarching focus to-date though is different from dealing with the inner most AI aspects of the self-driving car capabilities. Recently, GM announced that they would be digging deeper via the use of a private bounty program. Apparently, they have chosen a select group of perhaps ten or less white hat hackers that had earlier participated in the VDP and will now be getting a closer look into the inner sanctum. For aspects about GM's bounty program, see: http://www.autonews.com/article/20180803/OEM06/1808098 67/gm-hackers-cybersecurity-threats.

I've had AI developers ask me if they can possibly "get rich" by being a bounty hunter on AI self-driving cars. I wish that I could say yes, but the answer is a likely no. It might seem like an exciting effort of being a bounty hunter, wandering the hills looking for a suspect. It's not as easy as it seems. The odds of finding a bug is likely not so high, and how much you'd get paid is a key question too.

Consider too that you would need access to the AI self-driving car and its systems to even look for a bug. Right now, there aren't AI self-driving cars that are readily available on our roadways. Instead, the auto makers and tech firms are carefully watching over the AI self-driving cars that are on the public roadways. About the only means for you to get access would be to become a white hat hacker that gets invited into a private bounty hunter program for an auto maker or tech firm.

When the outlaw Jesse James was sought during the Old West, a "Wanted" poster was printed that offered a bounty of $5,000 for his capture (stating "dead or alive"). It was a rather massive sum of money at the time. One of his own gang members opted to shoot Jesse dead and collect the reward. I suppose that shows how effective a bounty can be.

Bounty programs have existed since at least the time of the Romans and thus we might surmise that they do work, having successfully endured as a practice over all of these years. For AI self-driving cars, I hope you will ponder carefully whether the use of a bounty program is worthwhile or not. The key overall aspect is that we don't want AI self-driving cars on our roadways that have bugs. I'll put up a Wanted poster right now for that goal.

CHAPTER 4
LANE SPLITTING AND
AI SELF-DRIVING CARS

CHAPTER 4

LANE SPLITTING AND
AI SELF-DRIVING CARS

Have you ever heard of lane splitting? How about the phrases of lane sharing, lane whitelining, or strip-riding? Here in California we are quite familiar with this terminology since it refers to something we see every day, namely, motorcyclists that go between the lanes on our highways, freeways, and byways.

This is one of those aspects about California that one might be conflicted about. We are the only state that specifically has made it a legal practice to do lane splitting. Some states outlaw it outright, while most states are silent on the matter and tend to either allow it implicitly or kind of look the other way about it. Is California right to have legalized the practice? Are we ahead of everyone else? Or, are we doing something unwise and unwarranted?

No matter where you are, I'd bet that there are rather divided opinions about the practice. The notion is that motorcyclists do not necessarily need to make the same kinds of lane changes that cars need to do. A motorcyclist is allowed under "lane splitting" to go between two cars and squeeze along forward.

Imagine you are driving your car on the freeway, doing so in the rightmost lane, and another car is to your immediate left in the fast lane that is adjacent to the rightmost lane. You and the other car are next to each other. You are both going say 50 miles per hour. You are both for the moment going the same speed and just a few feet from each other, really almost just inches at times. There is no chance of another car squeezing between you and the other car because you are both in unison and for the moment shoulder-to-shoulder of each other.

In California, a motorcyclist can try to squeeze between you and that other car. Imagine that the motorcyclist is doing 55 miles per hour and comes up from behind you and the other car. The motorcyclist is blocked seemingly because you and the other car are occupying the two lanes. Other cars behind you would need to wait until somehow an opening develops, such as if you speed-up or the other car does, and the two of you are no longer going neck-and-neck. Or, maybe you exit off the freeway and open up the rightmost lane for other traffic to proceed.

In any case, in spite of the apparent momentary blockade of you and the other car, if a motorcyclist believes they can fit between you two, they are legally allowed to do so. In fact, on most mornings as I drive on the congested bumper-to-bumper freeway to work, motorcyclists are streaming along and shimming between the cars. I can usually see them coming from behind me, snaking their way in and around cars, doing so while going perhaps 35 to 40 miles per hour, while the rest of the traffic is staggering along at maybe 15-20 miles per hour.

Presumably, the motorcyclists are polluting less than the car drivers and so they are being rewarded by being able to snake their way through the traffic. I know several motorcyclists that laugh when I tell them that my daily morning commute takes an hour or more. For them, by using lane splitting, they can do the same distance in half that time. I'd say they relish the aspect that the rest of the traffic is either sitting still or moving at a turtle's pace. They meanwhile are moving as fast as they can, albeit inhibited by the snarled car traffic.

You might be tempted to say that this practice seems reasonable. Why not let the motorcyclists be able to do lane splitting if it makes their commute more efficient? If you don't allow lane splitting, you would be forcing those motorcyclists to act like cars and be forced to wait behind the cars that are stacked up on the road. It would certainly be frustrating to the motorcyclist. Furthermore, from a traffic perspective, and especially here in crowded traffic-mania Southern California, having the motorcyclists act like cars would make our traffic even worse. For each motorcycle that might have zipped along and snaked through traffic, they would instead be taking up the same space as a car and cause our traffic lines to get even longer and generate more traffic congestion. Some also suggest that if motorcycles were relegated to staying behind cars while in-traffic, there would be more injuries or deaths of motorcyclists by cars that rear-end into the motorcyclists (another point of some debate).

I'd like to share with you a story about the sometimes-surprising nature of lane splitting. A colleague recently came out here from the east coast and he had never seen lane splitting in action. He had heard of it and had seen pictures and videos about it, but not had the "pleasure" of experiencing it directly. I decide to have some fun about this, and so I asked him to drive us to the office on his first day here. I wanted him to get a driver's view of the matter, rather than first being just a passenger in a car and experiencing the lane splitting from that seat.

For those of you that have not yet been driving when lane splitting occurs, it can be jarring when it first happens. My colleague was focused on the morning traffic mess and had forgotten my forewarnings about lane splitting. He was chatting with me and pointing out the crazy drivers up ahead of us, and then suddenly, seemingly out-of-the-blue, a motorcycle went past us, doing so within inches of the driver side window. It happened so fast that my colleague was startled and not even sure what had just happened.

All he saw was the flash out of the corner of his eye. He then turned his head in the direction of the motorcyclist, whom by now was already several cars ahead of us, having zipped past us and moving at a clip far above the speed of traffic. It was amazing to see the reaction of my east coast colleague. His mouth gapped open. He stammered that if he had opted to turn the wheel to the left and moved toward the left lane, he would have readily cut-off the motorcyclist and a dangerous incident might have occurred.

Welcome to Los Angeles, I said!

I mentioned that regrettably there are frequent such incidents of motorcyclist and cars that bop into each other. I would guess that at least once per week I witness one of these incidents or the aftermath of these lane splitting incidents. Personally, I think that's a frighteningly high frequency.

It has become commonplace for me to see a motorcyclist take a spill onto the freeway. Most of the time, thankfully, the motorcyclist gets back onto the motorcycle and continues on their way, apparently unhurt and undamaged. I see this repeatedly. On a few occasions, I've seen much worse, sadly. And, by listening to the traffic reports on the radio, every morning commute is filled with indications of motorcyclists downed here or there on our extensive freeway system.

In one sense, many seasoned drivers here take it for granted that there are lane splitting incidents. It is no different than expecting to see debris spilled onto the freeway by trucks that are overloaded or that have failed to cover-up their carrying loads. Each day, I see various debris such as fruits dropped onto the freeway, torn-up tires, old furniture, etc. I don't think it surprising. If I didn't see either flying or fallen debris on my daily commute, I'd be surprised. Likewise, if I didn't see a lane splitting incident over the course of a week, I'd wonder what happened that week (I was either asleep at the wheel or the motorcyclists decided to take the week off).

Thus, this takes us to the core of the controversy about the lane splitting approach. Some would say it is an overly dangerous practice and should be banned. Others say that it is up to the motorcyclists to decide what they want to do. If a motorcyclist is willing to take the risk, they should have the freedom to choose whether to do lane splitting or not. Just because it is legal here to do so does not mean that the motorcyclists all have to do lane splitting. They can use their judgement as to when it is safe to undertake it.

But, a counter-argument is that it is not just the motorcyclist that is involved in the lane splitting matter. By and large, a lane splitting incident is going to involve a car. Therefore, car drivers are just as involved. The car driver has no ability to decide when a motorcyclist should be allowed to do lane splitting. Since lane splitting is legal here, the driver must just live with whenever and however a motorcyclist decides to do lane splitting. Even if a motorcyclist does something really stupid and tries a lane split that is obviously ill-timed and going to likely produce a crash, there is no means for a car driver to particularly prevent it from happening.

Sure, you might argue that if the motorcyclist was in the wrong, presumably the driver of the car will be found to be not at fault and it will instead land on the head of the motorcyclist. Though this might be true, you need to factor into this the hassle part of the equation. The lane splitting has led to a car incident that otherwise would not have presumably occurred (at least not legally; i.e., of course a motorcyclist in a state that bans lane splitting could nonetheless do lane splitting, but at least they would already be in the wrong in doing so). And, the driver of the car will need to make their case that it was the motorcyclist that led to an incident.

Motorcyclists here will tell you that much of the time it is the "foolish" car drivers that are at fault. A motorcyclist trying to squeeze between two cars will often inadvertently get bashed by one of the two cars. A car driver might have swerved to the edge of their lane, doing so presumably without awareness of the presence of the motorcyclist. I've seen many sideview mirrors strike the motorcyclist, either damaging the mirror or tearing it apart from the car, and meanwhile it is usually enough of a blow that the motorcyclist loses control of their motorcycle and falls to the ground (by falling, I mean it is more akin to skidding along on the ground, since they are in motion at the time of the encounter).

The even worse encounters that I see involve a situation of two cars that are not actually abreast of each other, with one slightly ahead of the other, and for which one car suddenly decides to change lanes, and meanwhile the motorcyclist had mentally calculated that it was possible to squeeze between the two cars. The motorcyclist then rams into the rear of the car that opted to make the sudden lane change. This is worse than getting a glancing blow of a sideview mirror, and usually the motorcyclist goes down hard, often first having their body strike the back of the car.

There's another factor to these incidents that you need to consider. When a motorcyclist has an incident during lane splitting, it often inadvertently entangles other cars and car drivers into the incident.

Suppose you are driving along on the freeway at 35-40 miles per hour, and suddenly a lane splitting incident happens just ahead of you. Let's assume you weren't directly involved. But, you now have the chance of possibly running over the downed motorcyclist. Or, maybe you might hit their downed motorcycle. Or, maybe their motorcycle continues for a short distance, on its own, perhaps skidding, and strikes your car. Or, the car that was involved opts to slam on their brakes and has unexpectedly halted just in front of you. Etc.

You can be an innocent bystander that gets enmeshed in the whole mess that ensues.

Recently, the California Highway Patrol (CHP) announced new guidelines about lane splitting in California. Let's consider those guidelines.

First, the CHP recommends that only experienced motorcyclists try to do lane splitting. This is certainly sage advice. In reality, I'd suggest that most motorcyclists here on our freeways do lane splitting, regardless of their experience at riding a motorcycle. And, I'd also suggest that many of the newbie motorcyclists intentionally get a motorcycle partially because they are exasperated about waiting in traffic while driving a car. As such, they are determined to do lane splitting. It looks easy enough to do and likely these motorcyclists believe they can wiggle their way out of any trouble.

The CHP recommends that motorcyclists only do lane splitting when traveling at no more than 10 miles per hour faster on the motorcycle than the prevailing car traffic around them. Furthermore, the CHP recommends that lane splitting be only undertaken when the prevailing traffic is going below 30 miles per hour. More sage advice.

In my morning commute, I'd wager that the motorcyclists are often doing 20-30 miles per hour faster than the surrounding traffic when they lane split (rarely limiting themselves to just a 10-mph differential in speed). Also, the lane splitting seems to happen at all speeds, including when the prevailing traffic is going above the speed limit, such as 70-75 miles per hour.

I've seen lane splitting take place at speeds that make me shudder and for which if something goes awry it would definitely lead to death for the rider. At such high speeds, there is little or no room for error. The slightest twitch can spell the difference between a motorcyclist upright and one that is careening onto the asphalt and likely going to get hit by one or more cars along the way. We have a motorcycle helmet law in California, but I tend to doubt that a motorcyclist flying off their motorcycle at 80-mph and into fast moving traffic is going to have much luck even while wearing their helmet.

The CHP recommends that lane splitting only occur between lanes #1 and #2. This would be considered the fast lane and the lane to the right of the fast lane. We often have three, sometimes four, and even at times five lanes on our freeways in each direction. It certainly makes sense to suggest that the lane splitting happen only on the leftmost lanes. This keeps things a bit safer for the traffic in the slow lane and for when there are on-ramps and exits off the freeway. It would also provide a kind of consistency so that car drivers would know where to be on the watch for lane splitting.

Though this is again quite sage advice, I'd say that lane splitting seems to happen on any lane at any time. Whatever the traffic situation dictates and whenever a motorcyclist wants to get ahead of the traffic, there are those motorcyclists that do so. The aspect that especially seems dangerous and borderline legal is when they use the HOV lane for doing lane splitting. Here our HOV lanes are usually bounded by a double yellow that is not to be crossed at all, and only when there is a designated break in the HOV lane are you allowed as a driver to enter into or out of the HOV lane. Some areas of the country allow entry and exit of an HOV at any time. We generally do not.

The lane splitting motorcyclists will often slide back-and-forth into and out of the HOV lane. This often seems to work out well for those daring motorcyclists in terms of wanting to get ahead in the traffic. You might say that it maybe makes their efforts somewhat "safer" since the cars in the HOV are not supposed to be crossing the double yellow and likewise cars wanting into the HOV are not supposed to be crossing the double yellow. All I can say is that by my observations the lane splitting using the HOV seems to catch drivers especially off-guard and appears to be as dangerous, or more so than with lane splitting on the other lanes. Just an observation.

I'll cover just a few more of the CHP recommendations and not get to all of them. One recommendation is to not lane split near large vehicles such as buses and trucks. I agree wholeheartedly. Doing lane splitting in those situations is dicey since the driver of the oversized vehicle often cannot see the motorcyclist and also since other traffic can lose site of the motorcyclist due to the large vehicle's physical size too. As you might guess, I've observed lane splitting even when there are large vehicles nearby.

The CHP recommends that motorcyclists doing lane splitting should be wearing brightly colored clothing and gear. Sorry, I'd say that most of the motorcyclists that I see are usually wearing traditional oriented motorcyclist clothing consisting of black or brown leathered jackets, and rarely do they have any kind of especially high-visible clothing or gear on them. Until or if the motorcyclist culture somehow changes toward brightly colored attire, I'd say that this CHP recommendation is unlikely to be widely adopted.

There are even recommendations for car drivers.

The CHP indicates that car drivers should not try to impede a lane splitting activity. Allow me to explain.

There are some car drivers that don't like the lane splitting. As such, they will at times intentionally position their car to block a motorcyclist that is trying to perform a lane splitting action. This has included getting so close to another car that they are almost willing to have their car scrape against another car. Usually, these car drivers that are anti-lane splitting will be watching for an upcoming motorcyclist and then shift in the lane to the side of the motorcyclist, trying to sneakily narrow any space and thus discourage the lane splitting action.

This can be a quite dangerous cat-and-mouse kind of game. The motorcyclists that are seasoned as lane splitters know that some cars are trying to intentionally impede the lane splitting. As soon as a motorcyclist detects this possibility, they will try to outwit the car driver by faking to one side and going to the other side. I've witnessed a motorcyclist that nearly got jammed up by a car driver that seemed to be intent on preventing the lane splitting, and after the motorcyclist managed to burst just past the car, the motorcyclist then slammed his fist down on the hood of the car and sped away. It gets that crazy on the roads here.

The final piece of CHP sage advice that I'll mention herein is the aspect that per the CHP recommended practices that car drivers are to try and aid or enable lane splitting by shifting in their respective lane at the time of the lane splitting action. This to me seems the most questionable suggestion of the various points made. It is one thing to tell car drivers to not impede lane splitting, it's another idea altogether to have car drivers try to make it easier to do.

Why do I seemingly object to this notion of car drivers helping lane splitters? Here's why. Though on the surface of things it seems like a good idea to have cooperative drivers, and in fact many drivers do shift in their lanes to help provide more room for a lane splitting action, it can also get out-of-hand. I've seen some car drivers that in their desire to be generous to the motorcyclist, shifted so far in their lane that it scared other nearby cars. Those other car drivers did not realize what the intent of the shifting car driver was, and instead thought that the car was weaving or perhaps going to make a sudden move into the other lane (maybe the driver is drunk, maybe the driver has lost control of their car, etc.).

Because other drivers don't actually necessarily know that you are trying to help a lane splitter, it can create other adverse consequences. Those other cars can begin to move or swerve due to wanting to get away from your actions. This can then create a cascading series of such moves. It can be disruptive to traffic overall. It can make for tension and dangerous situations. I realize that the drivers doing this are trying to be good citizens, but unfortunately it has repercussions that I think even they at times are oblivious to.

You might also find of interest that motorcyclists sometimes thank car drivers that shift over in the lane to accommodate a lane splitting action. The most common form of thanking involves the motorcyclist waving their hand in a friendly gesture to the car driver, doing so as they get past the car. One small aside about the gesture. Since sometimes the motorcyclists use one finger to make a pissed-off gesture to cars that cut them off or don't help the lane splitting, I've seen some helpful drivers that get utterly confused to get a hand wave and mistakenly think the motorcyclist is irked at them, when in fact the motorcyclist is trying to show appreciation.

What a world we live in!

What does this have to do with AI self-driving cars?

At the Cybernetic AI Self-Driving Car Institute, we are developing AI software for self-driving cars. In our view, this also includes the ability of self-driving cars and the AI to be able to contend with lane splitting.

For many of the auto makers and tech firms that are making AI self-driving cars, the notion of dealing with lane splitting is quite low on their priority list. Indeed, they would tend to say it is an "edge" problem. An edge problem is one that is not at the core of the overall problem that you are trying to solve. You assume that an edge problem can be dealt with at a later time, after having first solved the core. For auto makers and tech firms, the core involves getting an AI self-driving car to work on our roadways in a rather bland and typical manner, after which they will deal with exceptions and so-called edge aspects.

When I've had discussions with them about this matter, they tend to point out that only California has a law that legally allows lane splitting. Why worry about a law that is only pertinent to one state? They are trying to make AI self-driving cars that work for anywhere in the United States and so it is "obviously" a rather limited concern when it is only lawful in just California.

I refute this idea that it is only limited to California. As mentioned earlier, many states allow it by not explicitly banning it. Plus, I assert that motorcyclists all across the country at times will do lane splitting, even in places where it is banned (I'd bet that most motorcyclists think it is a pretty low chance they would get nabbed for doing an illegal lane split, unless they did so brazenly and stupidly in front of a police car).

Lane splitting is actually quite popular in parts of Europe and in many Asian countries. If you are making an AI self-driving car, I'd suggest you ought to be considering how the self-driving car and the AI will cope in countries besides just the United States.

So, I tend to reject the idea that dealing with lane splitting is a rather narrow topic of only concern to California driving.

To further pursue this notion by some that lane splitting is a rarity and to be neglected for now, I'd like to first introduce the notion that there are varying levels of AI self-driving cars.

The topmost level is considered Level 5. A Level 5 self-driving car is one that is being driven by the AI and there is no human driver involved. For the design of Level 5 self-driving cars, the auto makers are even removing the gas pedal, brake pedal, and steering wheel, since those are contraptions used by human drivers. The Level 5 self-driving car is not being driven by a human and nor is there an expectation that a human driver will be present in the self-driving car. It's all on the shoulders of the AI to drive the car.

For self-driving cars less than a Level 5, there must be a human driver present in the car. The human driver is currently considered the responsible party for the acts of the car. The AI and the human driver are co-sharing the driving task. In spite of this co-sharing, the human is supposed to remain fully immersed into the driving task and be ready at all times to perform the driving task. I've repeatedly warned about the dangers of this co-sharing arrangement and predicted it will produce many untoward results.

Let's focus herein on the true Level 5 self-driving car. Much of the comments apply to the less than Level 5 self-driving cars too, but the fully autonomous AI self-driving car will receive the most attention in this discussion.

Here's the usual steps involved in the AI driving task:
- Sensor data collection and interpretation
- Sensor fusion
- Virtual world model updating
- AI action planning
- Car controls command issuance

There are some that suggest that even if lane splitting is worthy of consideration, you can delay worrying about it by instead just letting lane splitting happen for the time being. In other words, if a motorcyclist wants to lane split, let them go for it. The AI of the self-driving car presumably could care less that the motorcyclist is doing the lane splitting. No need to contend with the matter. Just let it happen. It's all on the shoulders of the motorcyclist.

If your view of a self-driving car is that it is sufficient for it to drive like a novice driver, I suppose there is some merit to this point about ignoring the lane splitting. I've seen novice drivers that are so overwhelmed with the driving task that the last thing they notice or care about involves motorcyclists that are lane splitting. The novice tends to assume that if the motorcyclist is doing lane splitting, the motorcyclist knows what they are doing. It would be as though a bumble bee has flown around your car. Let it do so.

I cringe at this belief. Are we really expecting that true Level 5 self-driving cars are to be driving on our roadways in the same manner as a novice driver? I hope not. If so, we'll all be in a lot of trouble.

As I've mentioned many times, a seasoned human driver knows how to drive a car in both proactive and defensive ways. They are on the watch for patterns of driving situations that alert them to take in-advance action. They manage to avoid accidents that might otherwise have occurred. I'm not saying that humans are flawless. I am just saying that we are more than just lucky that as a society we do not have more car accidents than we already have. I'd assert that human driving skills are an amazing aspect that generally keeps us relatively safe on our roads. It is a marvel to every day do my daily commute and I don't encounter accident upon accident and upon accident.

In short, I reject the idea that lane splitting can be "ignored" and also that even if being considered that it should somehow be placed at the back of the bus, as it were.

Here's a few reasons why being aware of lane splitting is an essential car driving skill.

First, be aware that many of the AI self-driving cars are initially going to be very rudimentary in their driving practices. They will tend to be skittish drivers. We've already seen that some AI self-driving car systems will only abide strictly by the speed limits and are quite civil in their behavior towards other cars. This has led to situations wherein an AI self-driving car kept waiting for other cars to go first, or frustrated other cars into making rash moves that then potentially led to an incident or potential incident.

I realize that some pundits of AI self-driving cars will say that there is nothing wrong with the AI being skittish and that the real problem is those pesky human drivers. Outlaw human drivers. Allow only AI self-driving cars. Problem solved. Those pundits conjure a utopian world in which there are only AI self-driving cars on the public roads.

Let's talk about reality. Currently there are about 250+ million conventional cars in the United States alone, and those cars are not going to magically disappear or become true Level 5 AI self-driving cars overnight. Indeed, the use of human driven cars will last for many years, likely many decades, and the advent of AI self-driving cars will occur while there are still human driven cars on the roads.

This is a crucial point since this means that the AI of self-driving cars needs to be able to contend with not just other AI self-driving cars, but also contend with human driven cars. It is easy to envision a simplistic and rather unrealistic world in which all AI self-driving cars are politely interacting with each other and being civil about roadway interactions. That's not what is going to be happening for the foreseeable future. AI self-driving cars and human driven cars will need to be able to cope with each other. Period.

We can also add to the reality list the aspect of motorcyclists. For those pundits that want to eliminate human driven cars, they would undoubtedly be aghast at the idea of still allowing human driven motorcycles. If the human drivers are gone, so would the human driven motorcyclists.

If we did indeed wave a magic wand and had only AI self-driving cars, and if there weren't any motorcycles at all, the lane splitting topic pretty much becomes irrelevant. If we allowed for AI driven motorcycles, which is an area of ongoing research, you'd likely need to contend with lane splitting. But, in that case, you'd presumably be able to have cooperative behavior between the AI's of the self-driving cars and the self-driving motorcycles. They would electronically communicate via V2V (vehicle to vehicle communication), and agreeably allow for lane splitting (in theory).

I'd say that realistically there will be lane splitting and it will occur during the advent of the emergence of AI self-driving cars. Furthermore, it will occur during the era of AI self-driving cars that are skittish.

How does skittishness play a role? Remember that I mentioned the story of my colleague that was surprised when a lane splitting action occurred? I had told him beforehand about lane splitting, and yet when it happened, he was surprised. Luckily, he did not do anything rash.

Suppose a skittish AI self-driving car suddenly has a lane splitting motorcycle that darts within inches of the self-driving car. What will the self-driving car do?

The AI might detect the motorcycle and assume that the motorcycle is on a path to hit the self-driving car. Perhaps the AI directs the self-driving car to make a rapid lane change to avoid the motorcycle. Or, takes some other action that is not quite expected or anticipated by other drivers and nor the lane splitting motorcyclist. This could spell trouble for the AI self-driving car, and the nearby traffic, and for the motorcyclist.

From a Machine Learning (ML) perspective, suppose the AI encounters this lane splitting a multitude of times, but has no context for why it is occurring.

What should the ML learn from it?

If each time it happens the ML opts to take a sudden evasive maneuver, it could be that the ML gradually accepts that this is the right way to deal with the matter. What the ML has "learned" is not necessarily the proper kind of driving action to take.

Overall, I'm suggesting that if AI self-driving cars are skittish or novice style drivers and cannot explicitly deal with lane splitting, they might get themselves into trouble, including possibly making rash driving moves that could lead to harm for any human occupants in the self-driving car, along with harm to other humans in nearby cars and for the motorcyclist too.

Further, for those that suggest that AI self-driving cars will just learn how to deal with lane splitting, doing so after some number of driving hours and experiencing lane splitting, I'm not convinced that what the AI learns will really be the most prudent driving practices related to lane splitting.

If we also embrace the recommendation of the CHP that car drivers should aid the lane splitting action, it seems especially unlikely that the AI is going to figure that out on its own. More likely is that the AI would figure out to defend against it, rather than to try and enable it.

In general, I'd vote that the AI be explicitly programmed or trained in dealing with lane splitting.

This also must include the realities of how motorcyclists really do lane splitting. If you were to try and setup the AI to believe that lane splitting will only happen on lanes #1 and #2, and only when the motorcycle is just 10 miles per hour faster than the prevailing traffic, and so on, you'd be creating a rather limited and potentially confused AI system when it had to deal with the real-world lane splitting activities.

The AI needs to also be able to contend with the reactions of human drivers.

As already mentioned, some human drivers will try to aid the lane splitting, while others will try to disrupt or prevent it. The AI self-driving car needs to be aware of those potential actions and be ready to deal with those other car drivers.

Let's pile on about the circumstances of lane splitting. It could happen during daylight when visibility is perfectly clear. Or, it can happen at nighttime when it is dark and hard to see the motorcyclist.

It can happen during dry weather when the roads are readily driven on, or it can happen in the pouring rain and the roads are slick and slippery. There could be just one motorcyclist trying to do a lane split, or there could be a multitude of motorcyclists doing so, all at once (I've seen this happen many times during my daily commute, namely motorcyclists riding together as a pack or team).

Lane splitting – is it a boon to our driving world, or is it a piranha that should be curtailed?

I'm sure the debate will be going on for a long time about the merits of lane splitting.

Meanwhile, it exists, and it happens. AI self-driving cars need to be ready for it. None of us want AI self-driving cars that by intent or by happenstance ram into a lane splitter or get involved in the aftermath of a lane split that has occurred to some other driver. Let's make sure that the AI is savvy about lane splitting. I'm not willing to split hairs on that.

CHAPTER 5
DRUNK DRIVERS VERSUS AI SELF-DRIVING CARS

CHAPTER 5

DRUNK DRIVERS
VERSUS AI SELF-DRIVING CARS

On a recent Friday night, after an evening event, I got onto the road and had some trepidation due to the aspect that I would be using locally popular freeways and highways for which drunken drivers also often used late at night (especially on Friday's and Saturday's). I debated whether to instead use other less traveled roads, perhaps being able to avoid those potential drunken drivers, but it would have added considerable time to my journey home and there was no guarantee that I still wouldn't encounter alcohol-impaired drivers.

You likely know that drunk drivers account for nearly one-third of all traffic-related deaths in the United States (per stats by the National Highway Traffic Safety Administration). The rule-of-thumb is that there's an alcohol driving related death every hour, based on averaging the number of such deaths over the course of a year.

You might not realize that annually there are more than 1 million drivers arrested for driving under the influence (DUI), which is an equally scary statistic (one can only wonder how many of those might have gotten into a deathly incident were they not arrested!). Of course, the one million drivers only represent those that were actually arrested and so presumably there would be many more that didn't get caught.

Some surveys indicate that there are perhaps more than 111 million instances of DUI "episodes" per year by U.S. adult drivers (this is based on self-reported indications by drivers). Though those episodes might not include actual driving related deaths, they likely include a significant number of driving fender benders, car or pedestrian sideswiping, frightening near-misses, and other dangerous mishaps.

Overall, you would be wise to be on the watch for drunk drivers. You should presumably always be mindful of drunk drivers both at night and day, though I'd suggest that you should have an especially heightened awareness during the times of the day that drunk drivers are most apt to be on the road, along with considering the days of the week, and any other aspects involving seasonality or special occasions. For example, getting onto the roads just after the end of the Superbowl game would count as a time to be especially wary of drunk drivers (a few too many brewskies while watching the game).

For my Friday night drive, it was nearing midnight and I knew it was the witching hour for many drivers that had been drowning their sorrows or partying it up at the bars after work and were giving up on going to the clubs (the next big-time risky spot would be 2:00 a.m. when the bars close-up). I usually try to arrange my schedule so that any late Friday night driving will be relatively close to home, but in this case, I'd attended an event that was about an hour from my house and so it would unfortunately provide plentiful chances for interacting with drunk drivers during the lengthy sojourn home.

I entered onto the freeway and for the first few minutes it was an uneventful drive. I had come onto the freeway into the slow lane and started to make my way toward the faster lanes, rather than sitting out the whole drive in the slow lane. As I tried to get into the lane just to the left of the slow lane, I noticed a car up ahead that I was approaching quite rapidly. My speed was a reasonable 55 miles per hour, the posted speed limit, while other traffic was tending to go a bit faster at that time of night. The car ahead of me was obviously moving at a much slower speed than the prevailing traffic.

As I got closer to the car, it was evident that the driver was moving along at about 35 miles per hour, which was nearly 20 miles per hour slower than the posted speed limit and probably 30+ miles per hour slower than the prevailing traffic. You might at first assume that maybe the driver was having troubles with their car. Perhaps it would explain the extremely slow speed on the freeway at that time. But, the driver was not in the slow lane and they didn't have their emergency flashers on.

I also observed that the driver would tap on their brake lights, doing so periodically. I looked up ahead of the car to see if maybe they were following some other car. There wasn't any car anywhere near the front of the turtle pacing car. The driver had the lane nearly to themselves. There they were, crawling along, late at night, doing so in the lane to the left of the slow lane, and occasionally pumping their brakes as though they thought they were already going too fast.

In my view, this driver was performing the driving task in a manner that suggested they might be a drunk driver. If this had been during daylight, I might have been less inclined to sway towards the drunk driving aspect, but it was the proper day of the week and the proper time of the night to believe that this could be a drunk driver. The driver seemed to be unaware of the true speed of prevailing traffic. They weren't trying to stay out of the way of traffic, which if they had been in the slow lane and had their emergency flashers on, we could all sympathize and assume perhaps there's some kind of car mechanical problems.

I realize you might at first be thinking that this driver wasn't harming anyone, so maybe it is harsh to say they are drunk driving. Well, I'd like to differ with the suggestion the driver wasn't harming anyone. Car after car was having to avoid the snail-paced car. Some cars would come right up to the bumper of the sluggish car and then dart into another lane. Some cars that wanted to move across lanes of traffic were at times having to jockey around to find an opening either just in front of the plodding car or just behind it.

Imagine if you had water flowing in a stream and placed a rock in the middle of the stream. The slowpoke car was causing the other cars to contort themselves into dealing with getting safely around this "rock" in the middle of the freeway.

I figured that it wouldn't be too long before some other car misjudged the situation and ended-up getting into a car wreck. This might involve hitting the plodding car. It might involve other cars hitting some other cars as they were trying to get around the plodding car, maybe inadvertently colliding with each other while trying to dance out of the way of the slow car. Perhaps the plodding car might even rear-end another car, which I suppose could happen if some innocent driver got in front of the slow car and maybe the slow car driver was so out-of-it they might ram into the other car.

There is also the chance that one drunken driver can sometimes meet-up with another drunken driver. As I watched the slowpoke driver, I realized that luckily the surrounding drivers seemed to be cognizant of the plodding car and were giving it relatively wide berth. Suppose though that some other drunken driver came along and was impaired in their driving capabilities. They might not be as attentive to the plodding car and be so careful to avoid it.

I opted to make my way gingerly around the slowpoke car and continue with my journey. I was trying to decide whether to call 911 and report the car, but in this circumstance, there wasn't anything demonstrably wrong per se. It was more of a hunch about the driver and the nature of the driving situation.

About ten minutes later, I encountered what seemed like another potential drunk driver, though the situation was quite different than the slowpoke instance. I was moving along at the speed of prevailing traffic, doing so in the fast lane. To my left was the HOV lane, which I couldn't use since I did not meet the necessary requirements for its use (such as needing to have 2 or more occupants in the car).

I saw a car up ahead that was entering into the freeway. The car came onto the freeway at a lightening like speed. I'd guess the driver might have been going 90 miles per hour or more. The driver then cut across all lanes of traffic and just narrowly missed me, having cut me off as the driver decided to go directly over into the HOV lane. It was illegal to enter into the HOV lane at the point that the driver did so (there are designed areas that you need to use to enter into and exit from the HOV lane).

I suppose you could say this was just a rude driver and one that apparently was willing to flaunt the law. They were driving at excessive speeds. They drove recklessly and had cut across multiple lanes of traffic, causing other cars to tap on their brakes to avoid hitting the interloper. Keeping in mind that it was a Friday night, late at night, I opted to consider this driver to also be a potential drunk driver.

As the driver rapidly gained distance from me, I could see that the driver was weaving into and out of the HOV lane. It was as though the driver could not steer the car in a straight-ahead manner. This led me to further deduce that it was likely a drunk driver. Little regard for other cars, could not drive straight, speeding and driving recklessly, and so on. It all added up.

Once again, you might criticize that I've pointed out what seemed to be another potential drunk driver. You might be thinking that the driver did nothing to harm anyone. If they want to speed, let them do so, and presumably the cops will eventually get them and give them a ticket.

I'd like to emphasize that the other cars on the freeway were quite taken aback by this driver. As mentioned earlier, some drivers had to tap their brakes or maneuver their cars to avoid hitting the rocketing car. I admittedly didn't see any crash take place, but I'd say it was a pretty strong bet that this driver was heading towards something untoward. Plus, given the high speed involved, whatever might occur was likely to involve really bad outcomes (at least injuries, more likely deaths).

Let's consider the types of driving actions that could be an indicator of a drunken driver:

- Driving too slowly for the roadway situation
- Driving too fast for the roadway situation
- Nearly hitting another car
- Cutting off another car
- Swerving across lanes needlessly
- Straddling a lane without apparent cause
- Taking wide turns rather than proper tight turns
- Driving onto the wrong side of the road
- Driving onto the shoulder of the road
- Driving in an emergency lane
- Nearly hitting a pedestrian, bicyclist, or motor cyclist
- Being too close to the car ahead of it
- Stopping when it seems unnecessary
- Rolling past stop signs
- Running a red light
- Other

I realize that those driving actions could be accounted for by some other reason beyond just drunk driving. As such, I am not saying that it is automatically an indicator of drunk driving if you happen to see a car do any of those specific actions.

You need to look at the context of the driving situation. Is the adverse act a seeming pattern of driving or was it a one-moment act and then didn't reoccur? Are there extenuating circumstances that could justify performing the act? Were more than one of the acts used in combination? What was the time of day and the road conditions?

What was the weather like? Etc.

Overall, those types of driving acts are telltale clues that can be used to try and guess whether there might be a drunk driver involved. It is useful to try and assess whether a driver is a drunk driver, since it tends to suggest that they will be dangerous in their driving efforts and you would be wise to then take extra precautions when near them.

As a seasoned driver, I try to be on the watch for drunk drivers and then take precautionary or "defensive" driving postures to reduce my risk and to hopefully also reduce the risks of others. Other innocents that come along might not be as alert or not yet have seen enough to detect a potential drunken driver.

What does this have to do with AI self-driving cars?

At the Cybernetic AI Self-Driving Car Institute, we are developing AI software for self-driving cars.

Allow me to elaborate.

I'd like to clarify and introduce the notion that there are varying levels of AI self-driving cars. The topmost level is considered Level 5. A Level 5 self-driving car is one that is being driven by the AI and there is no human driver involved. For the design of Level 5 self-driving cars, the auto makers are even removing the gas pedal, brake pedal, and steering wheel, since those are contraptions used by human drivers. The Level 5 self-driving car is not being driven by a human and nor is there an expectation that a human driver will be present in the self-driving car. It's all on the shoulders of the AI to drive the car.

For self-driving cars less than a Level 5, there must be a human driver present in the car. The human driver is currently considered the responsible party for the acts of the car. The AI and the human driver are co-sharing the driving task. In spite of this co-sharing, the human is supposed to remain fully immersed into the driving task and be ready at all times to perform the driving task. I've repeatedly warned about the dangers of this co-sharing arrangement and predicted it will produce many untoward results.

Let's focus herein on the true Level 5 self-driving car. Much of the comments apply to the less than Level 5 self-driving cars too, but the fully autonomous AI self-driving car will receive the most attention in this discussion.

Here's the usual steps involved in the AI driving task:
- Sensor data collection and interpretation
- Sensor fusion
- Virtual world model updating
- AI action planning
- Car controls command issuance

Another key aspect of AI self-driving cars is that they will be driving on our roadways in the midst of human driven cars too. There are some pundits of AI self-driving cars that continually refer to a utopian world in which there are only AI self-driving cars on the public roads. Currently there are about 250+ million conventional cars in the United States alone, and those cars are not going to magically disappear or become true Level 5 AI self-driving cars overnight.

Indeed, the use of human driven cars will last for many years, likely many decades, and the advent of AI self-driving cars will occur while there are still human driven cars on the roads. This is a crucial point since this means that the AI of self-driving cars needs to be able to contend with not just other AI self-driving cars, but also contend with human driven cars. It is easy to envision a simplistic and rather unrealistic world in which all AI self-driving cars are politely interacting with each other and being civil about roadway interactions. That's not what is going to be happening for the foreseeable future. AI self-driving cars and human driven cars will need to be able to cope with each other. Period.

Returning to the topic of human drunk drivers, some AI self-driving car pundits have said that there is no need to contend with human drunk drivers because there shouldn't be any human drivers allowed on the roadways. By exclusively having only AI self-driving cars on the public roads, you'd be able to eliminate the aspects of having to do deal with human drivers at all, regardless of whether those humans might be sober or drunk, since they would not be allowed to drive cars.

As mentioned, this is a rather crazy and at best naïve viewpoint about the real-world. In the real-world, we are going to have human drivers and we are going to have AI self-driving cars. The AI self-driving cars need to be able to mix with human driven cars. I suppose maybe fifty or one hundred years from now we might somehow have agreed to get rid of human driving, but in the meantime there's going to be a lot of human and AI driving going on.

Another perspective by some pundits is that we should restrict human driving to particular lanes or roads and have AI self-driving cars also be restricted to particular lanes or roads. The theory is that if you separate the two, meaning that you have AI self-driving cars driving on their designated roads and you have humans driving on their designated roads, you'll avoid any kind of contention between the AI driving and the human driving.

Again, this is not a particularly practical approach. There would be a substantive cost to set aside the lanes or roads for their appropriate designated kind of driver, whether the AI or the human, and the infrastructure costs would be relatively high to achieve this. It would also tend to imply that there are likely going to be some paths that will be a disadvantage to one or the other approach, suggesting that perhaps the AI might get roads that are going to be circuitous to get to where a passenger wants to go versus via human driven car there is a faster path (or, vice versa).

Some counter-argue that they are simply suggesting that there might be specialized lanes like HOV lanes, wherein the lane markings are established to indicate whether the lane is to be used by an AI self-driving car versus a human driven car. Though this might seem like a less costly and easier way to deal with separating the two, I point out that the nature of the separation is rather slim in this case of merely marking lanes.

If all you do is mark lanes by painting on the asphalt or using botts dots, there is still the significant chance of having the human driven cars mixing into the AI self-driving cars. In other words, just as human drivers regularly and readily violate HOV lanes when they aren't supposed to be using the HOV, it would be tempting to human drivers to go ahead and jump into the AI self-driving car lanes, if those lanes were moving faster than the human driven lanes or for any other reason that the human might decide to do so.

Furthermore, since we are focused herein on drunk drivers, it seems quite unlikely that a drunk driver is going to respect the AI self-driving car lanes and stay out of them. The drunk driver, in their inebriated state, will perhaps not even realize that the AI self-driving car lanes are only for AI self-driving cars. Or, the drunk driver might think it "fun" to proceed into the AI self-driving cars lanes, doing so due to their intoxicated state of mind.

You also have the rather practical matter of how to allow the AI self-driving cars to get into their lanes and how to let the human driven cars get into their lanes. For example, let's pretend that you opted to change the HOV lanes to become designated as AI self-driving car lanes. This seems easy to do, since the HOV lanes already are in existence in some areas. You just tell humans to stay out of those now AI self-driving car lanes, and you program the AI self-driving cars to stay out of the other lanes of traffic that are for human driven cars.

The difficulty with implementing this will be the transit aspects of how the AI self-driving cars can get into and out of the designated AI self-driving car lanes. Right now, human driven cars enter onto a freeway and make their way across numerous lanes of traffic to then reach the HOV lane. If you kept the same overall roadway infrastructure and merely designated those HOV lanes as AI self-driving car lanes, the only existing means for the AI self-driving cars to use those designated lanes involves traversing across the other lanes of traffic, which means that you once again have human driven cars and AI self-driving cars mixing together.

In short, I assert that we need to assume that AI self-driving cars will be mixing with human driven cars. And, those human driven cars might at times do some nutty driving, especially when a drunk driver is at the wheel of the car.

For those pundits that are willing to concede that there will be a mixture of human drivers and AI self-driving cars on our roadways, they sometimes will say that there is no need for the AI self-driving car to do anything special about the fact that there are human drivers. In essence, they suggest that if the AI self-driving car just follows the law and properly drives, it has no need to be concerned with drunk drivers.

I call this the head-in-the-sand approach to AI self-driving car driving.

Imagine if you had a novice teenage driver that you were training how to drive a car. You tell the teenager to generally ignore the rest of the traffic and just drive the roads in a legal manner. The novice dutifully follows your instructions and stops fully at stop signs, obeys intersection traffic signals, remains under the posted speed limits, etc.

Is the teenage novice now really ready to drive on our roads? Would you feel comfortable that the novice driver will be able to contend with the practical day-to-day dog-eat-dog world of driving a car? I would dare say that the novice is not yet ready. That teenage novice driver has to learn how to deal with the other human drivers that will do things beyond the norm of so-called proper and legal driving.

Most of the auto makers and tech firms are currently knee-deep in trying to get AI self-driving cars to simply drive along our roads in a manner somewhat akin to the novice teenager capability. Beyond that kind of driving, the auto makers and tech firms tend to consider anything else to be an edge problem. An edge problem is treated as a narrower use case and you presumably ignore it or postpone it until at some later time you have the resources and span of attention to take a closer look at it.

I'd assert that these simpleton AI self-driving cars are going to be unsatisfactory and until the edge problems also get solved there is either going to be limited use of those "mindless" kinds of AI self-driving cars, or worse still those AI self-driving cars will get themselves involved in various car-related incidents with other drivers and pedestrians, some of which will involve human injuries or deaths. Besides the danger to humans, it also could spell a large public backlash against the advent of AI self-driving cars, slowing down or possibly even halting progress on AI self-driving cars.

When I observed a suspected drunk driver on my Friday night journey, I did more than simply detect that the car might be driven by a drunk driver. I also took defensive driving measures in anticipation of what the drunk driver might do. That's something that a novice teenage driver is unlikely to be aware of, namely, they are not versed in how to detect a potential drunk driver and are equally unsure of what to do if they even spot one.

Thus, for AI self-driving cars, they should be AI equipped to detect whether human driven cars might be driven by a drunk driver. This involves observing how the nearby cars are behaving. I had noticed that a car was driving in the lane to the left of the slow lane, moving quite slower than prevailing traffic, and that was periodically tapping on its brakes. These are all telltale clues that the driver might be a drunk driver.

When I refer to a drunk driver, I don't want you to necessarily think that the AI self-driving car will be able to know or determine that a human driver is actually in a drunken state per se. I'm not suggesting that the AI will be able to any certainty determine the physical and mental state of the human driver that is driving a car nearby to the AI self-driving car. The AI is merely observing the driving behavior of the human driven car. It is then a logical inference that the human driver is somehow amiss, due to the driving behavior. The amiss nature of the human could be due to intoxication, or it could be that they are just emotionally distraught, or maybe are suffering from some other physical or mental aliment.

What the AI self-driving car mainly needs to ascertain is whether the human driver is driving in a "normal" human driving fashion or whether the human is driving in a more reckless manner. If the recklessness suggests drunken driving, there is an overall generic pattern of how drunken drivers tend to drive. By deducing that someone is a drunken driver, the AI can then potentially aptly predict what the human driver will do. As a result of those predictions, the AI self-driving car can and should take proper defensive and evasive actions to avoid a potential collision or other adverse consequence of that human driver.

Suppose you got into a ridesharing car that was an Uber or Lyft provided service, being driven by a human driver. It's Friday night, and as per my earlier example, imagine if the ridesharing driver gets onto the freeway and up ahead you see a car that is going slower than traffic and tapping on its brakes. I'd bet that you would look anxiously at your ridesharing driver and assume or hope that they notice this car up ahead and how it is driving. If you felt that the ridesharing driver was oblivious to the situation, you'd likely even say something to your ridesharing driver, cautioning them to be on the watch for that other car.

Let's now substitute AI for the ridesharing human driver. A true Level 5 self-driving car should be able to on its own detect those kinds of driving circumstances and automatically adjust to it. If you got into an AI self-driving car and it did not detect that suspected drunk driver, you'd be pretty worried that the AI was not up-to-par for fully handling the driving task.

I've had some pundits that say the human occupant could just tell or warn the AI, but this is nonsensical. The definition of a true AI self-driving car is that it can handle the driving task, and now you are wanting to carve out that the human occupants are supposed to be advising the AI about the driving task? This is letting the AI off-the-hook, so to speak. Plus, it is generally impractical. Suppose there isn't any human occupant in the AI self-driving car and the AI self-driving is simply driving to get to some destination – there's no human inside to help out the AI. Suppose too that only a child is in the AI self-driving car, are you expecting the child to be watching the traffic so as to make-up for the deficiency of the AI? I think not.

I'm not suggesting that human occupants or passengers won't be interacting with the AI of a self-driving car. Indeed, I am fully expecting that human occupants will be able to interact with the AI. But, this does not imply that the human occupants therefore have some kind of duty or responsibility to then watch the road for the AI. When we get into a taxi or cab today, we assume that the driver will be watching the road for us. We assume that the driver is savvy about the driving task. We don't expect that we as a passenger will need to aid the driver in the driving task.

As an aside, I realize that we all have at one time or another been in a ridesharing car and found ourselves forced into advising the human driver, perhaps out of concern that the ridesharing driver was not paying attention to the road or was otherwise unaware of something significant that could impact the safety of the driving journey.

Yes, this might well happen in the case of AI self-driving cars. But, it should not be the expected nature of the AI driving. In other words, if a human occupant is desirous of bringing something about the driving task to the attention of the AI, that's fine, and the AI should indeed consider such input, but it should not be that the AI can readily only successfully perform the driving task by a reliance upon a human occupant. That's not a true Level 5 self-driving car.

An AI self-driving car can be driving and not only observe potential drunken driving, but also share with other nearby AI self-driving cars what it is detecting on the roadway.

In my example of having seen the seemingly drunk driver that was driving slower than traffic and tapping their brake lights, the odds are that other nearby drivers were seeing the same thing.

Indeed, I noticed that other drivers were giving a rather wide berth to the suspected drunken driver. I was not able to readily communicate with those other nearby cars, but they likely also saw me maneuvering out of the way too. We all were potentially of a like awareness that this was a suspected drunk driver, and we were each taking precautionary measures accordingly.

In the case of AI self-driving cars, via V2V (vehicle-to-vehicle) electronic communications, the AI of one car could communicate with the AI of other nearby cars.

Imagine if on that Friday evening there were several AI self-driving cars that were near to the suspected drunk human driver. Those AI self-driving cars could have notified each other to be wary of the car. This could be quite handy such that if another AI self-driving car begins to enter onto the freeway, it could be forewarned even before being able to directly observe the drunk driving car that there is a potential drunk driver up ahead.

In addition to communication via V2V, there is also the possibility of V2I (vehicle to infrastructure) being used for this same kind of situation.

Increasingly, roadways are becoming "smart" by adding various computer capabilities and the use of V2I could allow for a roadway to communicate with the AI of self-driving cars.

The roadway might have detected that there was a driver moving at a much slower speed than prevailing traffic, and it could then send a broadcast to nearby AI self-driving cars to be on watch for this other car.

If AI self-driving cars are not versed in detecting, predicting, and acting to avoid potential drunk driven cars, the desire to see a dramatic reduction in drunk driving car accidents might not budge much. It has been hoped and stated that the advent of AI self-driving cars will get us toward zero fatalities, including making drunk driving accidents a thing of the past.

This though assumes a world of all and only AI self-driving cars.

In a world of mixed human drivers and AI self-driving cars, we could sadly end-up with even more drunk driving car injuries and deaths due to AI systems that are not adept at contending with human drunk drivers.

As mentioned in my Friday night example, the other human drivers purposely got out of the way of the drunk driver. Suppose these other cars were AI self-driving cars and they were not yet "edge" wised to the nature of human drunk drivers.

As such, those AI self-driving cars might get entangled into a drunk driver in a manner and frequency even higher than if there were only other human drivers on the roads.

Drunk driving is bad.

We all want to find ways to reduce or even eliminate the dangers of drunk driving. AI self-driving cars that are not able to contend with drunk drivers are going to potentially increase the risks of drunk driving incidents rather than reduce them.

There will be a trade-off between potential reductions in drunk driving fatalities because people are riding in AI self-driving cars which are not being driven then by drunk drivers, versus the instances of AI self-driving cars that get hit or come in contact with human drunk drivers and get into a fatality. The AI needs to be savvy about how to detect, predict, and out maneuver those human drunk drivers. This is more than just an edge or corner case.

CHAPTER 6

INTERNAL NAYSAYERS AND AI SELF-DRIVING CARS

CHAPTER 6
INTERNAL NAYSAYERS
AND
AI SELF-DRIVING CARS

Are you a naysayer at your firm?

I've been a leader and executive in many companies and invariably there are naysayers that you will encounter over the course of your work career. These internal naysayers can be both a curse and a blessing (or, if you prefer the alternative sequence, I'll say they can be a blessing or a curse), depending upon the circumstances and the nature of the naysaying involved.

The word "naysaying" tends to have a negative connotation to begin with and so I'd like to set the record straight that there can be positive naysaying and there can be negative naysaying. The positive naysaying is the type of naysaying that in a sense provides "constructive criticism," meaning that it likely is perhaps hard to hear or take, but that it has merit and should be given due consideration. I'll define negative naysaying as naysaying that is done without any merit per se and offers no particular redeeming value (it tends to be purely destructive rather than constructive).

It is difficult to readily know when naysaying is positive or negative.

One of the reasons that it can be confusing or bewildering to separate a positive naysaying from a negative naysaying is that the naysaying itself might be delivered in a manner that is untoward. If someone is a naysayer and presents their naysaying in a harsh or demeaning manner, those that are the brunt of the naysaying are quite likely to adversely react to the naysaying. Thus, the message itself gets lost in the manner of how the message was conveyed. This can be unfortunate since then the naysaying is like the proverbial throwing the baby out with the bathwater in that the actual useful parts of the naysaying are bound to be ignored or discarded.

In some companies, once someone starts down the path of being a naysayer they often get labeled as such.

This then tends to cull and permeate whatever they might have to say. In essence, once you've been stamped as a naysayer, whatever you do or convey will be likely seen in that light. You could claim the sky is blue and it would be perceived as yet another naysayer kind of remark. It usually doesn't take much to get labeled as a naysayer and it is an albatross that will be very hard to shed.

In fact, if the naysayer label is so stuck on you that no matter what you do it won't come unglued, you either need to live with it or consider going elsewhere (or, wait it out and hope that there is so much turnover in the firm that the new entrants won't know of and won't become tainted about your alleged naysayer status). What's tough too is that being branded as a naysayer internally can even follow you to other jobs, since often in the AI industry everyone knows everyone else, and the naysayer needs to decide whether they want such a label for throughout their career.

In rarer cases the naysayer label can become a badge of honor. If you were considered one of the first or one of the brave that came forth and naysaid about something that later was shown to be the case, you might be heralded for having come forward. Partially this depends upon how the initial naysaying took place. Partially it depends on the significance of the naysaid aspects. Unfortunately, often the original naysayer is left out in the cold, having been the one to start the firm toward some kind of awareness that ultimately was good for the company, but the naysayer has been earlier punished for their at-the-time act and there is little or no intention of overtly rewarding the naysayer now.

It can nearly always be dicey to be the bearer of bad news. Bad news is bad news. Whomever brings up the bad news will generally get associated with the bad news. It doesn't particularly matter that you weren't the "culprit" or that you otherwise had nothing to do with the bad news. Instead, people tend to anchor to the notion that the person reporting the bad news somehow has an afterglow of badness.

This is why trying to convey bad news is a tricky ordeal. Some will even attempt to pass-off the bad news to someone else to then will bring it up, in hopes of having the badness stick to that person instead. Or, one might try an anonymous method of communicating the bad news. Or, keep the bad news hidden. Or, attempt to coat the bad news with seemingly "good" news aspects and then hope that others will realize on their own that it is really just sugar-coated bad news (and, maybe the afterglow won't fall onto the person that brought the news to their attention).

I'd say that for most of the numerous software development teams that I've been involved with, invariably there is at least one or more naysayer on the team. Indeed, studies tend to show that tech specialists are more prone to cynicism and more likely to question orthodoxy then many other types of workers. This makes the chances of having naysayers involved in systems development projects a somewhat likely probability.

For those that aren't used to naysayers, or that have had problems in the past with naysayers, it can be a rude awakening to encounter naysayers that might seem actually helpful or constructive. The knee jerk reaction to the naysayer is that any naysaying is verboten and to be stymied. In a sense, some consider naysayers as being trouble makers, as being malcontents, and assert that those naysayers are disloyal and should be expunged.

My management philosophy is that I want to gauge whether the naysayers are providing value and offering positive naysaying. If the naysaying is solely of a disruptive nature and provides no value, and perhaps is being done for purposes of spite or other non-value aspects, it certainly then can be the case that the naysayer is not much more than a trouble maker per se. But, this is not an aspect that can be axiomatically assigned to someone.

There are those too that suggest the naysaying propensity is shifting and widening as the change in the age eras proceeds. For the millennials and the Gen Z, it is supposedly the case that they are more likely than prior generations to offer naysaying feedback. In prior generations, it was considered the case that firms are right to do what they do, and if you wanted to keep your job then you needed to just remain buttoned up and go with the flow.

The newer generations are supposedly more akin to want to work in a place that is doing the right things, and they will voice their concerns, and be willing to take a chance rather than be saddled with something that they believe is wrong, immoral, or otherwise not in keeping with their beliefs.

Many leaders don't realize that if they simply suppress all naysaying, they will likely lose potential feedback that could aid them in overcoming an issue now, or an issue that will arise further down-the-road. I've worked with numerous CEO's that would lash out at any naysayer, and for which the firm came to realize this, and as such the CEO became surrounded solely by "yes men" of the type that would never present any negative news whatsoever. This then led those CEO's towards at times a rather bleak end that would cause calamity for the firm.

Let's for the moment differentiate between being a naysayer versus the more legalistic whistleblower moniker. For purposes herein, the naysayer involves situations of things that are taking place that seem to be unwise or misguided, while in the case of a whistleblower we'll say that they involve aspects that are outright illegal. I realize that a naysayer can be or become a whistleblower, depending upon the circumstances and whether the aspects involved would be considered illegal or criminal in nature. For this discussion, the naysayer is someone focused on aspects considered ill-advised, rather than matters that go beyond that and extend into criminality.

What does this have to do with AI self-driving cars?

At the Cybernetic AI Self-Driving Car Institute, we are developing AI software for self-driving cars. I also frequently come in contact with AI developers that are doing likewise work at various auto makers and other tech firms.

At times, I've had those AI developers from other firms bring up to me concerns that they have about the firm they are presently working at and what the company is undertaking. These AI developers at times perceive there are unwise or misguided actions taking place but are not sure what they should do about it. You might say they are on the verge of being or becoming a naysayer and are unsure of what to do about it.

In some cases, they are already labeled as a naysayer and are unsure of what to do about it.

I've also had managers or leaders from such firms that tell me they have naysayers and from a leadership perspective they are unsure of what to do about it.

To provide some indication of what the naysaying is about in the context of AI self-driving cars, let's first establish some foundational aspects about AI self-driving cars.

There are varying levels of AI self-driving cars. The topmost level is considered Level 5. A Level 5 self-driving car is one that is being driven by the AI and there is no human driver involved. For the design of Level 5 self-driving cars, the auto makers are even removing the gas pedal, brake pedal, and steering wheel, since those are contraptions used by human drivers. The Level 5 self-driving car is not being driven by a human and nor is there an expectation that a human driver will be present in the self-driving car. It's all on the shoulders of the AI to drive the car.

For self-driving cars less than a Level 5, there must be a human driver present in the car. The human driver is currently considered the responsible party for the acts of the car. The AI and the human driver are co-sharing the driving task. In spite of this co-sharing, the human is supposed to remain fully immersed into the driving task and be ready at all times to perform the driving task. I've repeatedly warned about the dangers of this co-sharing arrangement and predicted it will produce many untoward results.

Let's focus herein on the true Level 5 self-driving car. Much of the comments apply to the less than Level 5 self-driving cars too, but the fully autonomous AI self-driving car will receive the most attention in this discussion.

Here's the usual steps involved in the AI driving task:
- Sensor data collection and interpretation
- Sensor fusion
- Virtual world model updating
- AI action planning
- Car controls command issuance

Another key aspect of AI self-driving cars is that they will be driving on our roadways in the midst of human driven cars too. There are some pundits of AI self-driving cars that continually refer to a utopian world in which there are only AI self-driving cars on the public roads. Currently there are about 250+ million conventional cars in the United States alone, and those cars are not going to magically disappear or become true Level 5 AI self-driving cars overnight.

Indeed, the use of human driven cars will last for many years, likely many decades, and the advent of AI self-driving cars will occur while there are still human driven cars on the roads. This is a crucial point since this means that the AI of self-driving cars needs to be able to contend with not just other AI self-driving cars, but also contend with human driven cars. It is easy to envision a simplistic and rather unrealistic world in which all AI self-driving cars are politely interacting with each other and being civil about roadway interactions. That's not what is going to be happening for the foreseeable future. AI self-driving cars and human driven cars will need to be able to cope with each other. Period.

Returning to the aspects of naysayers, there are a multitude of ways in which an AI self-driving car might be designed, developed, tested, and fielded, and therefore a lot of opportunity for disagreement about what is the appropriate way to do so. For auto makers and tech companies that are underway in developing AI self-driving cars, they each are taking a proprietary approach, involving making numerous assumptions about technology and about business and societal matters.

Let's consider some of the "naysayer" indications that have been brought to my attention.

- Naysayer raising concern that the company is focusing on the less than Level 5 rather than the Level 5 self-driving cars and will be unable or incapable of ultimately achieving a true Level 5 because of the misguided path underway.

- Naysayer upset that the firm is focusing inordinately on the use of LIDAR and that this is being done to the detriment of the other sensory capabilities.

- Naysayer that finds themselves always being brushed aside during team meetings going over the team's coding efforts and that there are qualms that the code is overly brittle and will not stand-up to the rigors needed to operate in a real-time on-board system.

- Naysayer expressing that there is insufficient time to test the systems being developed and that the assumption seems to be let the self-driving car in the wild discover any bugs and deal with it then, rather than trying to ask for more time now to do a more careful job of things.

- Naysayer worried that the Machine Learning (ML) and neural networks are being established with little or no attention to the potential of adversarial attacks and that this could be a potential hole that might be exploited later on by a bad actor.

- Naysayer indicating that some crucial assumptions about how pedestrians will abide by self-driving cars and about how human drivers will provide leeway to self-driving cars is shaping the AI in a very constrained and unrealistic way.

- Naysayer claiming that the AI developers are so overworked that they are tending to cut corners and are being treated like hamsters on a wheel that just need to shut-up, do the coding, and work around the clock to meet the stated deadlines.

- Naysayer that says the whole effort underway is a "cluster mess" (actually used a harsher four-letter word) and no one seems to be really in-charge and the whole AI system seems to be a hodgepodge which will either not work or work in unpredictable and dire ways.

If you simply read each of those naysayer remarks, I'd dare say that it is not readily feasible to assess whether or not the naysayer has a valid point or not. It could be that they are stating exactly what is taking place and that it bodes for rather serious and notable problems regarding the production of a safe and sufficient AI self-driving car. Or, it could be that the naysayer is jaded, perhaps having been passed over for some reason, and they are just lashing out for reasons of their own divine.

When I mention this aspect that without further context we cannot discern for sure what the situation really is, I usually get attacked right away by either the naysayer or those around the naysayer such as their supervisor or manager.

Lance, how can you possibly give credence to such preposterous claims, I am at times admonished.

Lance, you care about AI self-driving cars and how they will impact society and the public, so you must recognize that of course the remarks are true and must be immediately acted upon, I am sternly told.

There's no question that all of those naysayer remarks are worthy of getting attention. I am not suggesting that those remarks aren't worthy. As the proverb indicates, sometimes where there is smoke there is fire. Some of the remarks suggest that there might be really serious and damaging aspects that are not being taken into account for the AI self-driving car effort being undertaken. That's something to be given due attention.

I also want to mention the other side of the coin too. There are some AI developers that came from a university research lab or a governmental AI group that are unfamiliar with the ways of private industry and systems development therein. As such, they can sometimes get upset about practices that are different from what they experienced before. Those practices might or might not be valid. I'm just suggesting that the naysayer can be commingling the aspects of doing things differently than they are used to, along with the notion that what is being undertaken is "wrong" to do.

As such, I don't think we can axiomatically declare that any of those naysayer comments are right-or-wrong per se. Instead, it would be important to look into each matter and try to gauge what is taking place and how the comments potentially reveal actual aspects of concern.

If I am otherwise unable to get involved in exploring the naysayer remarks, I usually offer some overall suggestions of what the person might consider doing. Likewise, if a manager asks me about what to do about a naysayer, and assuming that I'm not able to further probe into the matter, I offer some overall advice. Let's get to those general pieces of advice and hopefully it might be of aid to you too.

I usually bring up that getting input from fellow colleagues of the naysayer can be helpful toward trying to ascertain the validity of the naysaying. Is the naysayer the only one that perceives things the way that they do? I'm not suggesting that the naysayer is therefore mistaken if they are the only one, since perhaps they are the only one that sees things more clearly than the rest. Also, the fellow colleagues might share the same beliefs as the naysayer, but be hesitant to state as such, due to concerns about losing their jobs or becoming labeled as a naysayer.

As such, it can be a difficult aspect to try and get open feedback from the naysayer colleagues. Any tech leader that just thinks to bring them all into a room and ask outright what's going on, well, it might work or it more likely probably would not work. There are many tech-oriented managers that are not especially versed in the subtilties of human behavior and are so focused on the tech side that they are unaware of how to actually interact with and manage people.

In theory, firms usually have an internal Human Resources (HR) department and any talent related matters should involve the HR team. This can be handy due to the HR team presumably being skilled in such matters, and also since there are potentially HR-related legal aspects that a tech leader or tech naysayer might unintentionally or intentionally violate in the process of pursuing naysayer claims. I realize that not all firms have an HR group, or they might have an HR group that is seen as not especially capable or maybe even arduous to deal with. Anyway, as mentioned, in theory there should be involvement by the HR team or someone versed in HR matters.

I also bring up that whomever is the immediate supervisor or "boss" of the naysayer should be consulted on the matter. Again, I'm not saying that the supervisor will be fully impartial and necessarily provide a reasoned analysis of the naysayer remarks. It could be that the supervisor and the naysayer are not seeing eye-to-eye, maybe on a lot of things far beyond the naysayer remarks, and so there is a bias potentially about anything the naysayer might have to indicate.

It is important to try and do some "egoless" analysis of the naysayer claims. For those of you that are developers, you might be aware of the notion of doing egoless reviews of system designs and coding. This involves looking at the design or the code and separating who did it from what it is. It used to be that when someone criticized or verbally "attacked" a design as being poor or the code as being bad, it was simultaneously an attack on the person that rendered the design or the code. The idea is to try and separate the artifact from the person and look at the artifact on its own merits.

This is not easy to do. It takes some very delicate and professed human behavioral skills to carry out an egoless design review or egoless code review. In any case, I am saying that you should try to separate the claims of the naysayer from the naysayer per se and look at the claims as a kind of artifact, similar to looking at a system design or lines of code. This might give you a chance at more realistically assessing the validity of the claims and whether or not the naysayer is offering some potential gold nuggets that otherwise might have gotten obscured by the rancor involved.

If trying to go the internal route is not viable regarding naysayer aspects, another approach can involve going external. Some firms arrange to have an external company that will for example take anonymous complaints on behalf of the firm (or, sometimes non-anonymous but with a promise of confidentiality), and then explore those for the firm. I realize that it is easy to be cynical about such arrangements. Maybe the outside firm has been told to simply bury any complaints. Maybe the outside firm will secretly report the complaints to the firm leadership and do nothing about it other than get you in trouble. Etc.

Another approach involves contacting an outside advisory group. Some firms for example have an alumni group of former employees. This might or might not be a means to try and get some useful advice (you never know and be forewarned that it could also open up a can of worms). There are also various industry professional associations that can sometimes aid in such matters (again, exercise due caution).

Going outside the firm can be problematic.

If you are bound by various confidentiality agreements, you might be violating those and thus are taking on more problems than perhaps you had anticipated. Sometimes too the act of going outside is instantly considered wrong by the firm, and so no matter whether your naysayer remarks are valid or not, you'll now be cast as a violator of company policy and procedures, for which your naysayer aspects will no longer count anymore, and it will become instead a focus on your wrongful action of going outside the firm. If you are considering going outside, likely wise to consult a qualified attorney, if you hadn't already yet done so.

Some naysayers become determined to seek out internal retribution come heck or high water. Even if they have AI skills that are in high-demand and they can step across the street to get a new job, they have become so enmeshed in the matter that they won't walk away from it. This can be out of a sense of duty, believing that they don't want to later on in life look back and think that they didn't do enough to stop what maybe later became a Titanic kind of disaster. Or, it can be that they are so emotionally wrapped up in the situation that they are willing to walk on hot coals to prove their points.

There are some AI developers that are aware of the bystander effect. This relates to the idea that sometimes bystanders will watch as something untoward unfolds and do nothing about it. You might be walking down the street and see a hoodlum that rushes up to an elderly person and steals their wallet or purse. Suppose that none of the nearby bystanders takes any action to prevent it or stop it. For those people that are conscientious and were bystanders, they will afterward be racked with guilt and anxiety that maybe they should have acted, maybe they should have tried to intervene.

For some naysayers, they are determined to not be part of a bystander effect. They might view that their cohorts are letting things happen that should not be happening.

These so motivated naysayers are willing to fight at preventing what they perceive as misguided or ill-advised efforts of developing an AI self-driving car. For those that are indeed going to prevent some later calamity, we can be thankful that they are willing to dare to speak-up. On the other hand, for naysayers that are without merit, and if their efforts merely delay and confound efforts toward AI self-driving cars, we would likely look askance at whether they are doing the right thing in their actions.

Naysayers, they can be a pain in the neck. Naysayers, they can be a live saver. You cannot just out of context declare a naysayer as one or the other of those types. A firm would be wise to put in place internal mechanisms to allow for naysayers to share their concerns. These mechanisms should be fair and balanced. Try to get value out of naysaying. Try to prevent subduing of naysaying for which it could have saved the firm and saved lives of those that might someday be a passenger in an AI self-driving car.

CHAPTER 7

DEBUGGING

AND

AI SELF-DRIVING CARS

CHAPTER 7

DEBUGGING

AND

AI SELF-DRIVING CARS

Do you know how the word "debugging" originated?

It is attributed to Grace Hopper, a pioneer in the computer field. When I was first starting out in computers, I won a programming contest and was lucky to have Grace Hopper hand me the trophy. I spoke with her that day, and indeed for the rest of my career it was my honor to keep in touch with her, along with heeding her sage advice along the way.

The story she told about the origins of the word "debugging" goes something like this. In the early days of computing, the rather massive-sized mainframe computers were using vacuum tubes as a means of memory. Vacuum tubes tended to generate a lot of heat. Grace was working late one evening and some of the windows were opened to let the heat vent to the nighttime sky. All of a sudden, the entire system came to a halt. She was tasked with investigating to figure out what had caused the system to stop.

Upon carefully inspecting the numerous hardware components and especially the vacuum tubes, she came upon one vacuum tube that had shorted out. There was a dead moth that had fluttered in via the open window and had landed on the vacuum tube at the hardware connection point, shorting out the circuit. Similar to a string of Christmas lights that goes dark if one bulb is bad, the shorted out vacuum tube had caused the entire mainframe to come to a halt. Grace removed the moth.

When she was asked what had happened and what she did to fix it, she said that she had "debugged" the system. For many years, the word "debugging" was an insider term that was used mainly by computer people. Eventually, the notion of debugging became popularized and we today use it for any kind of situation in which you fix a problem by removing or repairing some kind of error or bug. Thanks goes to Grace Hopper for adding the term to our global vocabulary!

The glory of programming always shines on the new development of code and tends to downplay or even ignore the nature of debugging a system.

As a former professor, I can attest to the aspect that my colleagues considered debugging to be something that students learning to program should just figure out on their own. It was considered something rather trivial and not worthy of any direct attention. There was a backward kind of logic often used by these colleagues, namely that if someone doing programming does the job "right" there should not be any need for debugging. Only foul-ups need to debug.

With my many years as an executive in software development organizations, I used to try and explain to them that debugging is actually a significant part of the job of professional programmers and software engineers. This idea that somehow code will be perfect is nonsensical.

Many of my academic colleagues had never worked on large-scale applications that consisted of hundreds of thousands or millions of lines of code, and thus for them a "large" program of a few hundred lines they envisioned should be written correctly and without any kind of debugging needed. They imagined programming as writing a mathematical proof.

The general rule-of-thumb in industry is that about one-third to maybe one-half of the development time of programmers or software engineers is consumed by doing debugging. Think about that statement for a moment. It means that relatively expensive labor is spending a significant chunk of their time and attention on debugging. If the amount of time consumed was say 1% or even perhaps 10%, we might be willing to just consider it as a minor aspect of the job, but when it rises to 30% to 50% of your time, it implies that it is substantive. Accordingly, it should be given due attention to try and ensure that it is as well spent as possible, minimizing when feasible the effort and providing suitable means to enhance it.

Oddly enough, in spite of those percentages, the world of system development is still generally in the dark ages when it comes to debugging. It still to this day tends to be the shadowy part of the job. Unspoken. Unheralded.

Here's then another perhaps startling figure for some, the total cost of a software development over its entire life cycle will usually come to about 50%-75% toward debugging. Once again, this highlights how important debugging is.

I realize that some of you will complain about the 50%-75% figure and point out that the bulk of that debugging comes after the system has already been fielded. The debugging at that juncture often occurs because something in the system environment has changed and so the code itself has to be changed too. Also, there are often new requirements that come up and the code written for those requirements can at times get lumped into "debugging," which seems like an unfair categorization. I'll grant you that all of that is the case, and so we might debate somewhat about what should be tossed into the "debugging" bucket, but nonetheless there's really not much question that debugging is a hefty proportion of the total life cycle cost.

I'd wager that most developers learn how to do debugging the old-fashioned way, simply by trial-and-error and learning it by the seat of their pants. No one likely explained to them the various techniques of debugging. Any of the tricks of debugging they had to discover on their own. They might have had someone more senior that gave them some pointers or aided them when doing some rather knotty debugging, which helped a little bit to get up the debugging learning curve. Overall, debugging is seen as a kind of apprenticeship skill and one that is considered more art than science.

Companies that provide specialized tools for debugging have often struggled to get developers to actively use the tools. Often, a developer will use such a debugging tool in a very rudimentary manner and avoid going to the effort to get into any more advanced debugging features. This lack of exploring those other debugging features might be due to a lack of awareness of what advantages they provide. There is also the aspect that no one is necessarily urging them to dive more deeply into being skilled at debugging. Most development managers assume it is just something that ripens as you get longer in the tooth in terms of developing more and more kinds of systems. By osmosis, the more you code, you just somehow presumably will get better at debugging.

Of course, one Catch-22 for many developers is that they are so pressured to debug systems that they don't have the time to focus on getting more effective and efficient at debugging. They just keep plowing away on fixing problems, and unfortunately over time there is not much advancement in their debugging other than becoming more familiar with a particular system that perhaps they are assigned to. Thus, the familiarity boosts their debugging rather than due to an increase in their actual debugging skillset.

Some will say that maybe I am glorifying the use of debugging tools and suggesting that debugging tools are a kind of silver bullet. No, I'm not in that camp. I know some that believe that we should be able to automate entirely the debugging process and, in a sense, take the developer out of the equation. There have been many attempts at applying AI to the debugging process. This involves infusing into the debugging tools the "mindset" of a developer.

I'd say we are about as far along on having AI that can generate new systems for us as we are on AI that can do debugging for us, which is to say that we aren't very far along on that path. Researchers that look at programming as nothing more than assembling Lego blocks have tried to come up with a means to automatically produce code and automatically debug code. Most of those efforts have fallen short by quite a bit. For now, systems development still seems to be a cognitively complex effort requiring human intelligence and skill. Maybe someday it will be more routinized, but not for the foreseeable future, I'd say.

Speaking of the mindset of a developer, studies of debugging tend to suggest that debugging is often greatly shaped by the mental model of the developer with respect to the system they are developing.

For example, a developer that originally developed a system will often have various assumptions in their mind about what the code is supposed to do. When the code does not perform as such, the developer can at first be quite flummoxed. This is why it is often the case that another developer, one that is not so close to the same code, can at times spot the issue, rather than the original developer, because the other developer does not necessarily carry the same assumption in their mind about the code.

Another element about debugging involves the use of language for developing a system. For most programming languages, there are a myriad of intricacies about what the language proposes a line of code will do, including whether you are using version X or version Y of the language, and whether the compiler or interpreter was established to execute the code in the way that you think it should. And so on. As a result, the programmer might have a mental model that the particular programming language they are using should do something M, when in fact at execution it does something else N.

One of my favorite such examples involves something so simple that it is wild to think that it still to this day stymies many programmers and many programming languages, and many development efforts. The simple matter of the initialization of variables is something that continues to be the exasperating culprit in many debugging scenarios. I might have written a line of code that says $J = G + 1$ and suppose that I had neglected to beforehand have something that gave G a value, such as I had meant to beforehand say that $G = 4$.

At execution time, if I had mistakenly not initialized G, the attempt to read the value of G and add one to it can do all sorts of crazy things. The system might assume that G was intended to be zero, and so at the execution time it decides that J will be calculated as though it says $J = 0 + 1$. Or, it could be that G is considered to have an undefined value that can be whatever value the system wishes to use, and so maybe the system decides at that moment in time that G is the value 31,727. Or, it could be that the value of G is considered a null, and the system maybe raises an exception or generates an interrupt that the calculation is invalid because the G is non-numeric. Etc.

The point being that our programming languages are so complex and brittle that there is bound to be a gap between what the developer thought would happen versus what really does happen. When I mentioned earlier that perhaps another developer upon looking at code, such as looking at the $J = G + 1$, might ask how does G get a value, and therefore find the "bug" that the original developer did not see, I don't want you to assume therefore that all we need to do is have a second developer look over the shoulder of the code of another developer. That's not a silver bullet either.

Some say that if we focus on the topic of debugging that we are then assuming that we are allowing bugs to be introduced to begin with, and we ought to instead be finding ways to prevent bugs from getting introduced. Well, I certainly agree with the part about wanting to try and avoid introducing bugs. Yes, I'm all for that.

Realistically, you need to realize that bugs are going to get introduced, and thus we need to look at debugging too. Don't fall into the simpleton argumentative trap of somehow thinking these are mutually exclusive aspects. You can be trying to find ways to prevent bugs from getting included, and you can also be trying to find ways to detect bugs that do get included.

One of my seminal articles on software engineering involved an analysis of the role of so-called software factories. There was a period of time that it was thought that development should be construed as an assembly line. Each member of the development team is to be considered on an assembly line and each has their portion of the assembly process assigned to them. The hope was that we could use the techniques found in factories and manufacturing to potentially improve software development, leading to faster development and less error prone systems.

In studying developers in the United States, I found that most developers tended to be mavericks, cowboys as it were (and/or cowgirls), and preferred to work independently, even when on an assigned team. The United States culture tended to consider development to be a solitary task. Meanwhile, in Japan, there was a movement toward the software factory under the belief that the team-oriented approach would be perhaps a better fit for a culture oriented more so to team kinds of efforts.

The twist I discovered, and which became a notable point at the time, was that even in the case of the software factory, human behavior still comes to play and needs to be considered. Allow me to explain.

The software factory approach assumed that if there was a bug found in the code, the bug should be traced back to the developer that originated it. Furthermore, it was thought that if the developer had allowed one bug to occur, the odds were that there were more bugs too. The developer was assumed to be a flawed process. The flawed process, i.e., the programmer, likely generated more bugs since inherently they are presumed to be flawed.

I realize we can all see the logic in that approach. Sure, if a programmer makes one error, maybe it would be the case that they have allowed or made other errors too. At the time, in the United States, once a bug was discovered and fixed, everyone went along on their merry way. In the case of the software factory approach, the developer that had introduced the bug was tasked with not only resolving the bug, but also had to spend a designated amount of time to relook at their overall code, doing so to find other potential hidden bugs. The thought was that there must be more bugs to be encountered, and so find them before they actually arise.

For some of the developers in the software factory, they did not like the idea of having to use time to search for other bugs that might be of their own causing. They would rather have shrugged off the one bug as a fluke and just kept going further on new development. But, the software factory approach required that they devote a set amount of time to find those other assumed bugs. Plus, if they could not find another bug that was originated by them, this was worrisome because the assumption was that at least one or more such bugs must exist. Presumably, it would be sitting out there, waiting to strike at an inopportune time.

As a result of this circumstance, some of the developers decided to play a little trick. They would purposely seed an innocuous "bug" into their code, doing so beforehand. When an unintended bug of theirs arose, they would first dutifully fix the unintended bug, and then rather than having to go on a witch hunt to find some new additional bug, they would wait a little bit and then announce they found a second bug (the innocuous one that they had purposely seeded). This would then satisfy everyone that the assumed other bug(s) had been found, and the developer would then be allowed to proceed on their other assigned tasks.

And so goes the nature of human behavior.

One of the key tenets of debugging that should be blatantly branded on the signboard of all developers is the idea that "first, do no harm." I mention this aspect because there is often a tendency to fix a bug and then inadvertently introduce an additional bug at the same time as enacting the fix. The programmer giveth and the programmer taketh, as they say. This is why debugging can be highly dangerous, since it is possible that you might fix a rather benign bug and find yourself accidentally having introduced a new and perhaps worse and more volatile bug.

I've had many situations wherein I faced making a leadership decision as to whether to have the developers fix a bug that was perhaps annoying but not incapacitating or take a chance on fixing the bug and maybe end-up with a much worse tornado on my hands. I used to take my car for repairs to a small auto shop that I had the same problem with. I'd take the car in for an oil change, and after I drove away the oil was fine but then the brakes weren't working well. It became a game of trying to decide how serious the existing problem was and what kind of other outcome they might produce. Yes, I eventually found a different auto shop.

Another key aspect of debugging involves being a kind of scientific detective. Here's what I mean.

When you are trying to figure out in a scientific inquiry what is taking place in a complex system such as the human body, you normally generate a hypothesis (notably, hypothesis generation is a crucial part of the scientific method). With your hypothesis in-hand, you then try to find evidence to either support the hypothesis or to disconfirm it.

For novices that do debugging, they often just start looking anywhere and everywhere to find what the bug might be. A more seasoned debugger will first craft a hypothesis. It might not be written down and might only be in their head. It might be well formulated and detailed, or it might be a hazy sketch. In any case, based on whatever they so far know about the bug, they craft a hypothesis. They then search for clues that allow them to refine the hypothesis.

This is what proficient developers do in order to try and more efficiently and effectively perform debugging. They perhaps weren't taught to do this, and just figured out this approach on their own, or watched someone else and opted to do the same approach as them.

Over time, their ability to craft on-target hypotheses gets better and better. You might liken this to a medical doctor that over time is more likely to be able to diagnose patient reported sicknesses, doing so because as a medical expert they have matured in their hypothesis generation and hypothesis testing capability.

What does this discussion of debugging have to do with AI self-driving cars?

At the Cybernetic AI Self-Driving Car Institute, we are developing AI software for self-driving cars. As such, we undertake debugging when needed, and we also confer with and advise other AI developers doing similar work for auto makers and other tech firms about ways to enhance their debugging skills.

The amount of software in an AI self-driving car can be enormous. Some estimates suggest that there might be on the order of 100+ million lines of code (LOC's) involved in the "typical" system for an AI self-driving car.

That's a lot of code.

You might be thinking that it is too large a number and that the number of human developers you would need would have to be enormous to produce that kind of volume of code.

Well, I hope you won't think it "cheating," but the number of LOC's includes not just hand-crafted code for various core aspects of the AI, but it also includes lots of other code too. For example, most AI self-driving cars are using various canned libraries of code and using open source provided code, etc. That LOC gets included in the overall count.

You also need to consider that the AI self-driving car has lots of sensors, including radar, sonic, cameras, LIDAR, and the rest. There is a lot of code needed for the internal sensor drivers that enact the sensor hardware and get those sensory devices to work. The count of the AI self-driving car code includes that code too.

There is also various Machine Learning (ML) elements, including artificial neural networks (ANN). In terms of counting that code, it is somewhat ambiguous about how to count that as the equivalent of conventional LOC's. Some try to count the underlying ANN execution on the basis of the LOC's used for that capability. Others consider the code used in the portion of the ML tools to do the training of the ANN. It's a bit open-ended about what is counted or not counted.

We'll also include in the count the rest of the automated systems for the car. There are systems devoted to the AI part of the self-driving car, but you also need to keep in mind that the car is still a car, and so it has a lot of other "traditional" kinds of systems for the ECU (Engine Control Unit) and all of the other various internal systems for the brakes, steering, etc.

One aspect you need to be aware of involves the aspect that there are varying levels of AI self-driving cars.

The topmost level is considered Level 5. A Level 5 self-driving car is one that is being driven by the AI and there is no human driver involved. For the design of Level 5 self-driving cars, the auto makers are even removing the gas pedal, brake pedal, and steering wheel, since those are contraptions used by human drivers. The Level 5 self-driving car is not being driven by a human and nor is there an expectation that a human driver will be present in the self-driving car. It's all on the shoulders of the AI to drive the car.

For self-driving cars less than a Level 5, there must be a human driver present in the car. The human driver is currently considered the responsible party for the acts of the car. The AI and the human driver are co-sharing the driving task. In spite of this co-sharing, the human is supposed to remain fully immersed into the driving task and be ready at all times to perform the driving task. I've repeatedly warned about the dangers of this co-sharing arrangement and predicted it will produce many untoward results.

Let's focus herein on the true Level 5 self-driving car. Much of the comments apply to the less than Level 5 self-driving cars too, but the fully autonomous AI self-driving car will receive the most attention in this discussion.

Here's the usual steps involved in the AI driving task:
- Sensor data collection and interpretation
- Sensor fusion
- Virtual world model updating
- AI action planning
- Car controls command issuance

Another key aspect of AI self-driving cars is that they will be driving on our roadways in the midst of human driven cars too. There are some pundits of AI self-driving cars that continually refer to a utopian world in which there are only AI self-driving cars on the public roads. Currently there are about 250+ million conventional cars in the United States alone, and those cars are not going to magically disappear or become true Level 5 AI self-driving cars overnight.

Indeed, the use of human driven cars will last for many years, likely many decades, and the advent of AI self-driving cars will occur while there are still human driven cars on the roads. This is a crucial point since this means that the AI of self-driving cars needs to be able to contend with not just other AI self-driving cars, but also contend with human driven cars. It is easy to envision a simplistic and rather unrealistic world in which all AI self-driving cars are politely interacting with each other and being civil about roadway interactions. That's not what is going to be happening for the foreseeable future. AI self-driving cars and human driven cars will need to be able to cope with each other. Period.

Returning to the debugging topic, trying to detect a bug in an AI self-driving car can be quite tricky. In addition, tracking down the bug is likely to be arduous. Plus, fixing the bug, and doing so without upsetting something else, well, it can be doubly tricky to do.

Why would it be hard to discover bugs in an AI self-driving car and be able to ferret them out and fix them?

You've got a myriad of subsystems that interact with each other. There's a wide variety of those subsystems in that some of them your own team might have actually written, or it might be subsystems that come along with the sensory devices or that come with the libraries used for your programming languages, etc.

Where throughout that overlapping, highly intersecting and convoluted set of subsystems upon subsystems should you look to find a bug?

Also toss into this equation that you are dealing with a real-time system. A bug that seemingly arises might actually be based on something else that took place seconds ago, or maybe minutes ago, and only after some time has elapsed does the bug "surface" to the attention of the overall system.

Suppose there is a bug in the radar sensory data collector software. On some periodic basis, the interpretation of the radar data is going to be incorrect. This is being fed into the sensor fusion. The sensor fusion is likely based on the assumption that what it is receiving from the radar software is considered "correct" and can be relied upon. Let's pretend that the radar is reporting an image of a structure up ahead in the roadway, but it is a bug in the radar subsystem and there is no structure there.

The sensor fusion might have been written to compare the radar with the vision processing subsystem that is using the cameras of the self-driving car. Let's assume that the cameras are working correctly, and the software indicates that there is not a structure up ahead. The sensor fusion now has to decide which is correct, the vision processing subsystem or the radar subsystem?

Let's pretend that the sensor fusion is written to assume that if either of the vision processing subsystem or the radar subsystem suspects a structure is ahead, it is "safest" to indicate to the virtual world model updating subsystem that there is indeed a structure ahead. The virtual world model subsystem is keeping track of what objects are around the self-driving car. So, it dutifully places a virtual marker of a structure into the virtual world model at the place that the radar says there is one.

The AI action planning subsystem examines the virtual world model and is trying to figure out what actions to have the AI self-driving car undertake. The AI now assumes that a structure is up ahead in the roadway, since it so noted in the latest updated virtual world model. As a result, the AI urgently issues car controls commands to have the self-driving car swerve around the structure.

Let's back-up for just a second or two. Suppose a human passenger in the AI self-driving car is quietly drinking their coffee and headed to work for the day. All of sudden, the AI self-driving car makes a crazy swerve, doing so for no apparent reason at all (a kind of "Crazy Ivan," which those of you Cold War buffs might know of). The human spills their cup of coffee. Maybe the human even gets a bit of whiplash from the rather sudden movement of the self-driving car.

Why did the AI instruct the self-driving car to make a swerving action when there was no apparent reason to do so?

Imagine that you were one of the developers of the AI self-driving car system and you got asked that question. I've obviously already told you that the issue arose with the radar subsystem having a bug in it but put that to the side for the moment. Suppose all that you knew was that a human occupant in your brand of AI self-driving car had reported to the auto maker or tech firm that the AI self-driving did a radical swerve for no apparent reason.

Where do you start to look to find whether there is a bug or not? As I say, the more seasoned developers will try to use any clues they can about the nature of the alleged bug to craft hypotheses and then use those hypotheses to try and dig into finding the bug.

What makes this harder too is that there's not going to be one developer that somehow magically knows the nature of the entire AI self-driving car. Instead, you'll have many, many developers that each knows some smaller part of the AI system and the rest of the self-driving car. Discovering the bug will require working with those other team members and collectively sharing and searching. Working in teams adds further complexity. Plus, you might have some rather acrimonious arguments about where to look and what subsystem might be the source of the bug (there is often a lot of finger pointing involved!).

You also need to consider the nature of the Operating System (OS) that is running the on-board hardware and whether it might have played a role in the nature of the bug, either being the source of the bug or perhaps either magnifying the bug or maybe even hiding the bug.

There are also going to be a quite a number of IoT devices included into an AI self-driving car, and those might play a role in either causing a bug or helping a bug to become exposed and activated. There's too the OTA (Over The Air) electronic communication aspects and possibly an interplay too with the V2V (vehicle-to-vehicle) communications.

Right now, the more advanced versions of AI self-driving cars are being kept relatively confined so that the developers can presumably (hopefully) find bugs now, before those AI self-driving cars are fully allowed into the wild. Thus, one method of finding the bugs involves restricted road trials on public roads, another involves doing testing on specialized road tracks, and another involves using extensive simulations.

There are some pundits of AI self-driving cars that seem to believe in a magical world that involves utterly flawless AI systems that are driving utterly flawless self-driving cars.

This is nonsense.

There are going to be bugs in the AI systems of self-driving cars. Face up to it. There will be recalls of parts of the AI systems and parts of the self-driving cars. Things will wear out. Maybe a parameter passed to a crucial API gets distorted and produces some cascading problem. Plus, it could be that the bug is intermittent and not readily reproducible. On and on.

Some say that the OTA will save us because via a quick electronic download you'll have a patch put in place for any bugs.

This belies the reality that first the bug has to arise and be reported.

Then, the bug has to be figured out and fully discovered.

Then, a fix or patch has to be devised and tested, and presumably be tested such that it does not introduce some other adverse consequence.

All of that will take time.

Meanwhile, the AI self-driving car will presumably be driving along, and possibly have a bug that endangers the occupants, or jeopardizes humans in other cars or perhaps imperils nearby pedestrians.

Allow me to say that I am not an alarmist and proclaiming that the sky is falling. I'm simply trying to exhort all fellow AI self-driving car developers to realize the importance of debugging.

An AI self-driving car is a life-or-death real-time system. We need to realize that it is not going to be bug free. We need to be well-prepared for debugging. Lives will depend upon it. I would also predict that if we have too many bugs at the first foray of AI self-driving cars, it will spoil the barrel and we'll all experience a downturn in the support for and belief in AI self-driving cars.

We all must of course do whatever can be done to prevent bugs, but we must also face reality and be optimized and efficient and effective to cope with after-the-fact bugs and do an astounding job in finding and fixing them. Please don't let debugging be perceived as a last resort and something of little standing. Debugging is vital to the safety of humans that will be using AI self-driving cars, and crucial to the entire future of AI self-driving cars.

Debugging, make sure to put it on the top of your priority list.

CHAPTER 8
ETHICS REVIEW BOARDS
AND
AI SELF-DRIVING CARS

CHAPTER 8

ETHICS REVIEW BOARDS AND AI SELF-DRIVING CARS

As a driver of a car, you are continually making judgments that involve life-or-death matters. We don't tend to think explicitly about this aspect of driving and take it for granted most of the time. Whenever there is a car accident, the topic comes up about what the driver did or did not do, and any aspects of how judgment came to play in the accident usually comes to light.

Suppose you are driving down a street at nighttime. You have your radio on. It has been a long hard day at work and you are heading home for the evening. How well are you paying attention to the driving task? Perhaps your thoughts are focused on a difficult problem at work that you are hopeful of solving. The radio is meanwhile tuned to a talk show and it covers a topic of keen interest to you.

You normally take the main highway to get home, but tonight you opted to use a less common road that you hope has little traffic and will allow you to get home faster. The speed limit is 45 miles per hour, and you are doing about 55 mph. Going over the speed limit on this particular road happens all the time and going just 10 mph over the speed limit is actually not much of an excess in comparison to what other drivers do.

Suddenly, via your headlight beams, you see what might be a figure in the road up ahead. There's not a crosswalk nearby and so you weren't anticipating that any pedestrians would be in the roadway. You weren't looking for pedestrians, plus with your thoughts on the problems at work and with your somewhat rapt listening to the radio talk show, it all added up that you didn't notice the shadowy figure at first.

Your mind races as to whether it really is a person or not. The roadside lighting in this area is rather poor. You have only a few seconds of time to decide what to do. Should you slam on the brakes? But, if so, there is a car behind you and they might ram into your car. Plus, perhaps by slamming on the brakes you might lose control of the car and not be able to maneuver it. You could instead try to swing wide, out of your lane, and do so in a somewhat frantic manner under the belief that the shadowy figure is headed in the other direction. You might just skirt the figure by going to the left, if you can swerve just enough and if the shadowy figure continues to move to the right.

Swinging over into the other lane isn't so easy though. There is another car in that lane. You might cause the other driver to react and they might then swerve into the median. You could maybe try to go to your right, up on the sidewalk, doing so to avoid the figure in the street. But, it is so dark that you aren't sure if there might be anyone on the sidewalk and besides the idea of driving on the sidewalk seems almost crazy, really just a desperate last resort to avoid hitting the figure in the street.

This is a relatively realistic scenario and one that any of us could encounter.

Let's analyze the situation.

The driver is faced with a rather untoward dilemma. There might or might not be a pedestrian in the path of their car. Whatever is in the path, the driver only has a few seconds to decide what it is and what action to take.

If the driver opts to use their brakes, it could lead to the car behind the driver doing a rear-ender and it might injure or kill the human occupants in either or both cars.

If the driver opts to swing into the next lane to the left, it could lead to the car in that lane becoming concerned and possibly veering into the median, which could injure or kill the human occupants, and might careen further into traffic and injuring or killing other humans in nearby cars.

If the driver opts to drive up onto the sidewalk in hopes of avoiding the figure in the street, there might be pedestrians there that could get injured or killed, plus the driver might generally lose control of the car and the driver gets injured or killed too.

If the driver decides to stay the course and continue forward, they will potentially hit the shadowy figure. This might injure or kill the figure, assuming it is a human, and the driver might also get injured or killed in the process of striking the figure.

Is there a proper and precise equation or some form of calculus that we can use to identify what the correct course of action is?

I don't think so.

Suppose you had time to try and develop some kind of calculation, what would it consist of? You might try to find out the ages of the various "participants" such as the driver of the car, the driver of the car behind the dilemma facing car, the driver of the car in the next lane over, etc. Maybe you could say that the older the driver the more they have lived their lives and so the less they count in terms of whether to be on the one that might take the brunt of the situation. In other words, you might say that moving into the lane to the left is the "better" option because the driver in that car is the oldest of those involved and thus has already lived their life.

Some would say that your use of age in this manner is outrageous and absolutely wrong. You might instead try to calculate the societal value of each participant, somehow trying to encompass what they do and how they are helping our society. Or, maybe you come up with some other factors to try and weigh the value of their human lives.

You might instead just decide to use probabilities regarding the various actions involved. If the approach of slamming your brakes has a 30% chance of injury or death, while if you swing into the next lane there is a 60% chance of injury or death, perhaps you should go with the brakes option since it has the lower probability of an adverse outcome.

These analytic methods could be handy and yet it seems rather trying that any of us could individually come up with an agreeable set of equations or formulas to cover such circumstances for us and others. As far as we all know, the method used by today's human drivers is the nebulous notion of "human judgment." None of us can really say whether our brains do some kind of mathematical calculation, nor can we explain directly why we did something. We can rationalize what we did by offering an explanation, but the explanation itself might have little to do with what really happened inside our heads.

Explanations are provided as a means to try and turn our mental aspects into something that can be elaborated to other people. Usually, our explanations are intended to suggest a logical means of how we arrived at a decision. No one can though definitively say or prove that their mind actually carried out the logical steps offered. Instead, the explanation is a post-reflected aspect that might match to what our minds considered, or it might be a completely concocted aspect.

Suppose the driver in this case decides to go ahead run into the shadowy figure. Did they do so after carefully considering all of the other options?

The driver might after-the-fact claim they considered the various options, but perhaps they did and maybe they did not. It could be that the after-the-fact explanation is an attempt to rationalize what took place. The driver might not want to seem as though they just mindlessly rammed into the shadowy figure, and as such, provide instead an elaborated indication of the other options, which might allude to the notion that the driver tried to find a means to avoid the incident, even though maybe they just froze-up or maybe didn't even notice the shadowy figure beforehand at all.

Ponder for a moment the number of times that each of us as car drivers make these kinds of spur of the moment decisions, doing so in real-time, in order to try and avoid causing some kind of car incident that might injure or kill others. It's not just limited to those occasions when you get into a car accident. You undoubtedly have lots of situations that fortunately don't lead to an accident per se, and yet you had to make some tough decisions anyway.

In this case of the driver, suppose it turns out that the shadowy figure was actually a large tumbleweed that was blowing across the street. If the driver opted to plow ahead and into the tumbleweed, perhaps it led to no car accident. The driver just kept going. Meanwhile, the car behind also kept going, and the car in the lane to the left kept going. None of them are injured or killed. Yet, there was a split second or so when a decision might have been made that could have led to their injury or death. No one would have likely recorded this non-event and no explanation or rationalization was sought or tendered.

I'd like to suggest that with the millions of cars on our roads on a daily basis, we are all involved in millions upon millions of such judgment calls, continually, and those of us in the cars, either as drivers or passengers, are subject to the outcomes of those judgments. So too are the pedestrians nearby to wherever cars are driving.

It is actually a bit staggering that we don't have more car accidents. With this many people and they are all making those millions upon millions of judgements, it is somewhat a miracle that their judgments are good enough and sound enough that we don't experience even more car incidents and more injuries and deaths accordingly.

I hope this doesn't scare you from getting into your car. Also, I hope that this discussion hasn't been overly macabre or ghastly. As I suggested earlier, the reliance on human judgement permeates our car driving and determines life-and-death matters. We don't usually overtly consider this aspect in our daily driving and tend to take it in stride.

What does this have to do with AI self-driving cars?

At the Cybernetic AI Self-Driving Car Institute, we are developing AI software for self-driving cars. One crucial aspect to the AI of self-driving cars is the need for the AI to make "judgments" about driving situations, ones that involve life-and-death matters.

I've had some AI developers tell me that there isn't a need for the AI to make such judgements. When I ask why the AI does not need to do so, the answer is that the AI won't get itself into such predicaments.

I am flabbergasted that someone could have such a belief. In the scenario that I just described, I would assert that the AI could readily have gotten itself into exactly the same predicament that I had indicated the human driver was involved in.

Some might say that the AI would not be distracted by the radio playing and would not be thinking about problems at work. Okay, let's subtract that entirely from the scenario. Some might say that the self-driving car would not be driving over the speed limit. I'd tend to debate that aspect, but anyway, let's go ahead and assume that the self-driving car was doing the 45-mph speed limit.

We still have the situation of the car approaching the shadowy figure and need to consider the matter of car behind the self-driving car and the car that is to the left of the self-driving car, all being done in real-time, with just a few seconds to decide, and with the balance of people's lives at stake.

If you were to suggest that the self-driving car would be better able to detect the shadowy figure because the self-driving car has not only cameras but also radar, sonic, and perhaps LIDAR capabilities, I'd say that yes there is a chance of having a more robust indication, but in practical terms those sensors won't guarantee you that you have a better detection. Anyone that knows much about those sensors would concede that you can still have an imperfect indication of what is ahead of the self-driving car. There are many factors that can limit the capabilities of those sensors.

Some would say that the self-driving car would make sure to have sufficient distance between it and other cars so that it could have the needed stopping distance unimpeded. I don't quite see how that is feasible per se. If the car behind you is on your tail, how do you ensure that there is sufficient stopping distance without getting rear-ended by that other car?

The answer usually is that the other car is being driven by a human and the "stupid" human has not allowed for the proper stopping distance. Therefore, the problem now is that we have a human driver, which if we just remove all of the pesky human drivers and have only AI self-driving cars, we would not need to be concerned with cars being too close on our tails.

This will require me to take you on a related tangent about the nature of self-driving cars.

There are varying levels of AI self-driving cars. The topmost level is considered Level 5. A Level 5 self-driving car is one that is being driven by the AI and there is no human driver involved. For the design of Level 5 self-driving cars, the auto makers are even removing the gas pedal, brake pedal, and steering wheel, since those are contraptions used by human drivers. The Level 5 self-driving car is not being driven by a human and nor is there an expectation that a human driver will be present in the self-driving car. It's all on the shoulders of the AI to drive the car.

For self-driving cars less than a Level 5, there must be a human driver present in the car. The human driver is currently considered the responsible party for the acts of the car. The AI and the human driver are co-sharing the driving task. In spite of this co-sharing, the human is supposed to remain fully immersed into the driving task and be ready at all times to perform the driving task. I've repeatedly warned about the dangers of this co-sharing arrangement and predicted it will produce many untoward results.

Let's focus herein on the true Level 5 self-driving car. Much of the comments apply to the less than Level 5 self-driving cars too, but the fully autonomous AI self-driving car will receive the most attention in this discussion.

Here's the usual steps involved in the AI driving task:
- Sensor data collection and interpretation
- Sensor fusion
- Virtual world model updating
- AI action planning
- Car controls command issuance

Another key aspect of AI self-driving cars is that they will be driving on our roadways in the midst of human driven cars too. There are some pundits of AI self-driving cars that continually refer to a utopian world in which there are only AI self-driving cars on the public roads. Currently there are about 250+ million conventional cars in the United States alone, and those cars are not going to magically disappear or become true Level 5 AI self-driving cars overnight.

Indeed, the use of human driven cars will last for many years, likely many decades, and the advent of AI self-driving cars will occur while there are still human driven cars on the roads. This is a crucial point since this means that the AI of self-driving cars needs to be able to contend with not just other AI self-driving cars, but also contend with human driven cars. It is easy to envision a simplistic and rather unrealistic world in which all AI self-driving cars are politely interacting with each other and being civil about roadway interactions. That's not what is going to be happening for the foreseeable future. AI self-driving cars and human driven cars will need to be able to cope with each other. Period.

Returning then to the matter at hand of the scenario about the driver and the shadowy figure in the roadway, we need to dispense with the notion that the cars around the self-driving car will be only AI self-driving cars. Realistically, there will be a mix of human driven cars and AI self-driving cars.

I say this to clarify that the scenario I've painted remains the same, namely the AI is faced with the matter of having to try and determine whether to hit the brakes but might get rear-ended, or swing into the next lane but might cause the other driver to veer into the median, or the AI might drive onto the sidewalk but maybe harm pedestrians, or the AI might continue straight ahead and potentially plow into the shadowy figure.

As mentioned, there are AI developers that claim that an AI self-driving car would not let itself get into such a predicament, but there doesn't seem to be any realistic world in which the AI could have magically avoided this situation and many other such situations. I'm putting a stake in the ground and will unabashedly say that there are going to be unavoidable crashes that AI self-driving cars will need to confront (and, of course, there will be avoidable crashes too, for which hopefully the AI will be astute enough to avoid).

I've stated many times that there are crucial ethical decisions or judgments that the AI will need to make when driving a self-driving car. I don't believe you can hide behind the matter by saying that the AI will never get itself into a situation involving an ethical decision or judgment. Saying this belies the very act of driving a car. Anyone developing an AI self-driving car that seems to think that the AI won't get itself mired into such situations has their head in the sand, and worse too they are developing an AI system that cannot presumably handle the real-world driving tasks that the AI will face.

For the moment, please go with me on the notion that the AI will need to cope with ethical decisions or judgments as part of the driving task. If that is indeed the case that the AI will need to deal with the matter, the question then becomes how it will do so.

You might suggest that the AI needs to use common sense reasoning.

As humans, we seem to have an ability of being able to use common sense about the world around us. We somehow know that a chair is a chair and that the sky is blue. We also presumably use common sense to decide when to slam on our brakes in the car versus swerving into another lane. Well, sad to report that we don't yet have any true semblance of common-sense reasoning for AI systems, and so let's count out for now the "solution" that we could just plug-in common sense reasoning and have dealt with the ethical choices matter swiftly by doing so.

You might say that the AI should use Machine Learning (ML) to figure out how to cope with these ethics related decisions. Are you suggesting that we let AI self-driving cars drive around and sometimes they hit and injure or kill someone, and sometimes they don't, and by the collection of such driving instances that somehow over time the ML "learns" which approach to take in these dicey situations? This seems impracticable. I would wager that most of us would not want to be one of the humans injured or killed during the thousands of such instances that the ML needed to collect to be able to find patterns and "learn" from the experiences.

In short, the better approach would be to explicitly design, develop, test, and field the ethical decision making or judgment aspects into the AI.

Thus, since we don't have available as yet any kind of automated common-sense reasoning, and since relying upon ML to somehow miraculously over time figure out what to do (during which grave results are apt to occur), it would seem prudent to overtly tackle the problem and devise a system capability for the AI to rely upon.

If we do nothing, the AI will be unable to adequately perform when such moments arise, and the result will be likely random chances of the self-driving car either managing to avoid an incident or getting involved in an incident and doing so without any explainable rhyme or reason for it. I don't think we want self-driving cars to become clueless rockets of potential destruction.

Now, assuming that indeed the appropriate approach would be to devise a system component for this purpose of ethical decision making, this raises a slew of technological and societal considerations.

Should this be left to the auto makers and tech firms to devise on their own, each independently creating such system components? This would seem somewhat questionable. If you have brand X self-driving car driving around and it is going to decide one way as to how to ascertain whether to proceed forward toward the potential pedestrian or weave or hit the brakes, and there is brand Y self-driving car that decides another way, it would be potentially confusing for the public at large as to what to expect from the AI of these self-driving cars.

Besides the aspect that each of the auto makers or tech firms would need to reinvent the wheel, as it were, in terms of trying to come up with a viable approach, it would seem more consistent and transparent if some overarching approach were used. This too would deal with the potential thorny aspect that involves the crux of how the decisions are being made.

The thorny aspect involves how to decide what the "best" course of action might be in these ethical dilemmas. I had earlier asked whether humans use some kind of mental calculus to determine which choice to make. Do humans weigh each factor? Do they consider whether age is important of those that might get injured or killed? How do humans do this? We can't say for sure how humans do it.

This makes trying to have an AI system do something similar a problematic issue. It would be handy to know how humans make such decisions and thus we could just pattern the AI to do the same. I've had some AI developers that tell me that all this will take is to ask people how they decide, and then essentially "program" this into the system. As pointed out earlier, the rationalizations that people provide are not necessarily how they truly decide, and we are not even close as yet to being able to probe into the mind to discover how people really make such decisions.

Perhaps this takes us toward the ML approach and the need to collect sufficient data, though doing so via car accidents themselves would seem dubious. Another approach would be the use of simulations and have humans that gauge and make choices in the car driving simulations, out of which the ML might "learn" the approach being used by humans (even if we don't know what's actually happening in their minds).

Another approach would involve using an actuary's kind of analytics method. As emotionally difficult as it might seem, there might well be a need to identify and agree to factors that should come to play in these decision moments. The result would be developed as part of the AI for use in the on-board system of the self-driving car. The same kind of gut-wrenching aspects are involved in trying to decide actuarial matters and thus it seems potentially fitting to use the same kind of methods for these purposes.

Rather than leaving this task to the auto makers or tech firms alone, some have proposed that an Ethics Review Board mechanism should be utilized. These would presumably be special committees or boards that would meet to aid in determining the parameters and thresholds for use in the ethics aspect components of the AI self-driving car systems. It might be something crafted by industry or it might be something created via potential regulations and regulatory bodies.

These Ethics Review Boards might be established at a federal level and/or a state level. They would be tasked with the rather daunting and solemn task of trying to guide how the AI should be established for these tough decision-making moments (providing the policies and procedures, rather than somehow "coding" such aspects). They might also be involved in assessing incidents involving AI self-driving cars that appear to go outside the scope of what was already established, and thus be an ongoing aid in the re-adjustment and calibration of the implemented approach.

Some have suggested that if there was an AI component for these ethical decision-making moments, and if there is a desire to standardize it across self-driving cars, perhaps the component should be housed in the cloud. Similar to how self-driving cars will be using OTA (Over The Air) electronic connections to update the AI systems, perhaps the AI component would not be embedded into the on-board system of the self-driving car and instead be accessed remotely.

Of course, the remote access aspects might get in the way of the decision making itself. It is more than likely that the ethics component would need to be accessed in real-time with split seconds to render a choice. Doing so via electronic connection seems dicey and prone to being inaccessible at the moment that the aspect is urgently needed.

What would seem prudent would be to have an on-board capability that could be updated via a cloud or centrally based standard. The on-board component would then be honed to presumably be able to render a choice in whatever sufficient time is available in a given circumstance. If insufficient time existed in any particular instance, there would need to be some shortcut choice capability, which I mention since once again the thought is to avoid an arbitrary choice and one-way-or-another have a "reasoned" choice that can be understood and explained.

One question that some have posed is whether this ethics decision making component could be truly able to handle all of the many variants of the Trolley problem. For example, I've outlined the case of the driver that is not sure if a pedestrian is in the road, and there is a car immediately behind, and there is a car to their left. Surely there are thousands of such potential instances, all of which would have variants. How could a system possibly contend with so many variations?

I'll bring us back to the aspect that humans seem to be able to contend with these multitude of variants. I'd guess that the driver faced with the situation I've outlined has not experienced that exact situation before. Instead, they have an overall experience base and need to use whatever they can to try and apply it to the moment and the situation at hand. Presumably, the AI component would need to do the same. Plus, the AI component would be sought to be adjusted and enhanced over time (via the use of the Ethics Review Boards).

There is unquestionably bound to be controversy about the notion of the Ethics Review Boards. Some suggest that they should be called Safety Review Boards or perhaps Institutional Review Boards, providing a naming that might be more palatable. There are some that have pointed out that there is the possibility of having them become labeled as "death panels" as per the political term that arose during the 2009 debates about aspects of the federal healthcare legislation (this phrasing seemed to strike a chord with the public at large, though there is quite a dispute about the merits of the labeling).

In one sense, it could be argued that the Ethics Review Boards would be shaping how the AI will respond to dicey driving incidents, and as such, those Boards are deciding how life-or-death decisions will be made. It would be no easy matter for the members to serve on such a committee. Careful selection and criteria for participation would need to be figured out.

As unseemly as it might seem to have such Boards, the alternatives are to allow whatever happens to just happen or allow for particular auto makers or tech firms to make those a priori choices for us all. It would seem to be the case that society would likely prefer the more open and transparent and collective approach of using the Boards, but this is something yet to be ascertained.

A few final comments that I'd like to cover on this topic encompass various security related aspects.

One concern would be that a hacker might somehow be able to mess with the on-board ethics choice component and alter it so that it would do something untoward. When the ethics component is involved in a dire situation, the hacked version might make a choice that purposely seeks to maximize injury and death, rather than minimize it. Of course, systems security does need to be paramount for the on-board AI, and in fact I'd suggest that if the hacker could hack pretty much any part of the AI of the self-driving car, the odds are they can produce an untoward result in some fashion.

In essence, rather than focusing on solely the ethics component, nearly any other element of the AI system if hacked can likewise produce adverse consequences. As such, all I'm saying is that you cannot argue that there should not be an ethics component due to the potential for it being hackable, since you could make the same argument for nearly all other components of the AI system for a self-driving car. If you then are making the argument that any of those components could be hacked and therefore they are inherently untrustworthy, you might as well then say that there is no such viable thing as an AI self-driving car.

In a somewhat similar manner, let's consider the cloud and the OTA. One might argue that suppose a hacker gets to the cloud version of the centralized ethics component and messes with it. The hacker has made things presumably easier for themselves in that they didn't need to try and access any particular self-driving car, and instead they will let the OTA do so for them. The OTA would presumably blindly and dutifully send the updates to the on-board AI systems and thus allow a viral-like spread of the untoward ethics component aspects.

I'll invoke the same argument as before. Yes, if a hacker could hack the centralized version, it would potentially produce this kind of calamity. I would submit though that if the hacker could alter nearly any aspect of the centralized patches that are going to be pushed down to the AI self-driving cars, you can have an untoward result. As such, the security needs to be quite tight at both the on-board self-driving car and at the OTA cloud-based elements. Either one can allow for something untoward if the security is not sufficiently tight.

I've had some AI developers tell me that their "solution" to these ethical choice situations involves having as a default that if the self-driving car cannot decide what to do, it will simply slow down and come to a halt. I hope that you can readily see that such an approach is nonsensical. Using my earlier example, would we have wanted the driver to have simply slowed down and come to a halt? This is quite impractical in the given situation and as I say is a nonsensical way of thinking.

Another idea that has been offered would be to ask the humans in the self-driving car as to what the AI should do. Again, a nonsensical answer. First, suppose there aren't any human occupants in the AI self-driving car at the time of such a decision-making moment? We are going to have AI self-driving cars driving around on their own, quite a bit.

Second, even if there is a human on-board, would they be able to out-of-the-blue be able to make such a decision? Let's assume they aren't driving and aren't paying attention to the driving task, which in a Level 5 self-driving car is indeed their prerogative.

Third, suppose the human on-board is drunk? Suppose the human on-board is a child? Suppose there are humans on-board and yet the decision needs to be made within 2 seconds – how could the humans be told the problem and offer an answer in a mere second's worth of time. And so on.

Another point some make is that maybe we should setup remote human operators that would make these decisions. Sorry, it's a nonsensical idea. Suppose the remote operator could not fully grasp the nature of the situation? Suppose they only had two seconds to decide and meanwhile they somehow needed to "review" what the situation is and what options to consider. Suppose there are electronic communication delays or snafus and the remote operator is not able to participate in the time needed? And so on.

I'd say that the automation is what is going to get us into this predicament, and it would seem like the automation is the only means to get out of it (as coupled with the Boards and the approach to devising the solution).

Though, when I say get out of it, let's be clear that however this is devised, the odds are that the AI system will be second-guessed about the choices made. This would be true of humans and it will certainly be the same about the AI. The AI might "perfectly" execute whatever the AI ethics component consists of, and yet still human lives might be lost.

There are unavoidable crashes that no matter what you do, a crash is going to occur.

For my earlier example, suppose it really was a pedestrian in the roadway. And, suppose that each of the choices involved either injuring or killing someone, either the pedestrian, or the driver in the car behind you, or the driver in the car to your left, or you as the driver. There is not going to be a magical way to get out of the unavoidable crashes unscathed.

Would we prefer as a society to pretend it won't happen and then wait and see. Or, would we rather step-up to the matter and address is head-on. Time will tell.

.

CHAPTER 9
ROAD DIETS
AND
AI SELF-DRIVING CARS

CHAPTER 9
ROAD DIETS AND
AI SELF-DRIVING CARS

There must be hundreds or even maybe thousands of different diet regimens that you can opt to use. Some diets might be good for you, while others can have adverse consequences that outweigh the benefits of the dietary aspects. People often struggle to choose the right diet for them, and they equally tend to struggle trying to stay on the diet and stick with it. I've known some people that became quite irritable and onerous once they got onto a diet program.

Let's consider another kind of diet, namely a road diet.

If you aren't familiar with the notion of a road diet, don't feel too bad about not knowing what it is. In some sense, road diets are a bit of a fad that seems to come and go. Generally, a road diet consists of taking an existing roadway and altering it to reduce the number of traffic lanes or otherwise adjusting the nature of the traffic lanes. The roadway is usually still the same overall width. The width of the lanes within the overall width of the roadway are the target of the changes or adjustments.

Besides referring to these kinds of changes as a road diet, there are some that simply call it "lane reductions" (but that's not very catchy, is it), and others that refer to it as road re-channelization (a hefty 5-dollar word that makes it sound more scientific). To those that want to make lane reductions, using the phrase "road diet" is handy since it has a rather positive connotation. We all generally believe that diets are a good thing. Those that oppose road diets are a bit chagrined that the road diet moniker is used and claim that it is a sneaky wording that hides the true intent, consisting of reduce the number of car traffic lanes. In any case, I'll use herein the phrase road diet.

Allow me to provide you with an indication of what a road diet might consist of.

Suppose that your town or city has a four-lane road that is known as Main Street. There are two lanes going in the southbound direction and two other lanes going in the northbound direction. Traffic moves along on this handy thoroughfare. Perhaps this Main Street has been in existence for many years and seemed to serve the needs of the town or city quite well over those many years. It has become a heartened part of the tradition and lore of the place.

But, there are some in the community that have qualms about the layout of Main Street. It is dangerous for bike riders to ride on Main Street, and even when hugging the curb, there have sadly been periodic instances of car accidents involving wayward automobiles striking kids and adults on bikes. Another concern is that the cars driving on Main Street tend to go faster than the speed limit, often acting like they are driving on a four-lane open highway instead of down a busy street with lots of shops and businesses. There have been many circumstances of pedestrians that almost got hit while trying to cross Main Street.

What to do?

Some would say that this Main Street is primed to go on a road diet.

Here's what we'll do. The total width of Main Street is 44 feet. There are four lanes of 11 feet each. Let's get rid of two of those car traffic lanes, which frees up 22 feet. Since we want to help out the bike riders, let's use 5 feet respectively on either side of Main Street for a devoted bike lane. In the middle of Main Street, we'll put a new lane that's 12 feet wide and allows for making left turns.

Overall, here's what we originally had for the 44-foot wide Main Street:

- 11-foot southbound car lane
- 11-foot southbound car lane
- 11-foot northbound car lane
- 11-foot northbound car lane

Once we put the roadway onto the devised road diet, it would contain this:

- 5-foot bike-lane southbound
- 11-foot car-lane southbound
- 12-foot car-lane mix for south/north left-turning traffic
- 11-foot car-lane northbound
- 5-foot bike-lane northbound

We are still within the original 44 feet of Main Street. This makes life somewhat easier because if we had wanted to widen Main Street it would have been a quite extensive and expensive roadway infrastructure project. We would have had to uproot the sidewalks and the various fire hydrants and light posts. Since we are only changing the lanes within the existing overall width of the road, the effort to make the changes will be a lot less costly and arduous to undertake.

I'm not suggesting that the road diet changes are somehow cost-free or cheap to do. Having to re-stripe the road and potentially make other modifications to accommodate the new plan can definitely have some substantive costs. If you though compare those costs to widening the road or making other more substantive structural changes, on a relative basis the road diet is likely more affordable.

Not all road diets will necessarily stick with the original width of the road. You can have instances of widening the overall width of the road, even when it is undertaking a so-called diet. The diet part of things is usually focused on the fact that you are reducing the number of car traffic lanes (or, sometimes reducing the width of the existing car traffic lanes). You then use the freed-up space for other purposes, which might include adding bike lanes, adding a center turn lane, or perhaps widening or adding sidewalks, etc.

For the Main Street example, suppose the changes were indeed made and we have this new and exciting version of Main Street. Bike riders are now less likely (hopefully) to get hit by cars because of the added bike lanes. This might encourage bike riders to use Main Street, more so than they might have otherwise.

Constricting the car traffic to now just one lane in each direction is likely to slow down the cars. This might deal with the prior aspect that cars were often speeding down Main Street, rising to speeds that were prone to accidents and potentially hitting pedestrians. With the constricted availability of just one lane in each direction, the traffic is perhaps slowed down and will be less apt to drive recklessly.

There are some studies that claim that the proper use of a road diet can reduce car crashes by around 47% and reduce speeding by about 70%. Those studies also suggest that there might be an increase in bike riders of around 37%. Plus, there might be an increase in pedestrian foot-traffic of around 49%.

It would seem that the use of a road diet is a really good way to keep a roadway in existence and yet redesign and re-purpose it to better suit the needs of the community. It can potentially save lives. It can possibly rejuvenate an area – suppose that the bike riders and pedestrians had previously avoided going to the businesses and shops on Main Street due to the car traffic dangers. Now, those bike riders and pedestrians might opt to revisit Main Street and shop there once again.

Not all is necessarily so rosy on Main Street, though.

If the drop to just one car traffic lane in each direction does constrict traffic flow, it might lead to heavy congestion now on Main Street. Cars might be snarled all along Main Street, trying to get to their destinations. This heavy traffic might become visually a blight and might also increase noise or odors. It could frustrate car drivers. People might now find themselves taking much longer to drive to wherever they are trying to go and thus burning up more gasoline and wear-and-tear on their cars.

An unintended reaction by the drivers could be that they decide to spill over into nearby neighborhoods. If Main Street has become a bottleneck, the car drivers might decide to turn onto side streets and weave through whatever streets are adjacent to Main Street. This could impact those streets and endanger those that live on those streets. All of a sudden, a quiet neighborhood that once had only local traffic could now be inundated with cars trying to make their way along on Main Street but desperately seeking an alternative.

This self-diverting of traffic could increase the risks for pedestrians and bike riders in the adjacent neighborhoods. Whereas maybe Main Street is now safer, it could be that the risks and potential car accidents are merely being shifted into those other streets. You didn't particularly fix the problem and only pushed it into another area. Sometimes the adjacent neighborhoods now need to react and ask that roadway speed bumps be put in place, along with other traffic restrictions and posted signs to warn car drivers to not carelessly use those adjacent streets as though they are still on Main Street.

Some would argue that another disadvantage of constricting the car traffic involves the potential delaying of first responders to an emergency. A police car that is trying to quickly get to a crime scene and that uses Main Street might be delayed by the traffic congestion now on Main Street. Likewise, there might be delays to ambulances or fire trucks. This could be another unintended consequence of the road diet.

Another possibility of something amiss could be that cars begin to avoid using Main Street whatsoever. This certainly reduces the volume of traffic and might aid the use of Main Street for the bike riders and pedestrians. But, it could also lead to less people driving to and visiting the shops and business that are lined along Main Street. Those shops and businesses might soon discover that their revenues are drying up due to the road diet.

A somewhat rarer potential issue could be that circumstance of a mass evacuation and the road diet preventing people from readily driving to get out-of-town. If a hurricane is heading in the direction of the town and people are supposed to get out-of-town, perhaps the reduced lanes might slow down all that car traffic and prevent people from fleeing on a timely basis.

As you can hopefully discern, the road diet is a practice that often involves great controversy.

Many cities or towns that start toward using a road diet approach will often do so quietly and without much fanfare. It kind of slips under the radar of the populace. A particular town might decide that they have a roadway they want to put on a road diet. They move forward doing so. Once they are done, all of a sudden, the car traffic that has routinely been using that roadway now becomes quite concerned about what has taken place. People rise up and complain.

It could be that the car traffic that was using that portion of the roadway as a kind of pass-thru and they really didn't care much about the local aspects per se. If the pass-thru was in a location that does not allow for ready alternatives, it is likely those car drivers are going to be steamed about the changes. The potential political pressure and public backlash can be tremendous.

Plus, even if those impacted are sympathetic to the road diet, they might argue that the particular road diet design was deficient. As I mentioned earlier, people that sometimes choose to go on a food diet discover that not all food diet programs are necessarily the best for you. It might be that the road diet approach undertaken is not the best choice for the situation at-hand and thus it could be that some that oppose the road diet are primarily opposing the specific implementation of it, and yet still open to a road diet of some alternative design.

One such example of a road diet controversy took place in Santa Monica, California in 2017, and I was caught up in the matter as someone that at the time routinely drove through the area in question. Press coverage encompassed both those in favor of the road diet and those in opposition. The opposing forces called it a draconian lane reduction, some called it a debacle, some said it was a road diet disaster. There were even efforts to recall some of the politicians that had been involved in the road diet effort. As you might guess, there was a lot of hand wringing too because some that generally believe in road diets were worried that if this road diet was expunged it might curtail all future road diet initiatives.

Generally, there is much debate on all sides of the road diet approach.

I mentioned earlier that it could be that the road diet might slow down first responders, but this is considered a controversial point and there are some researchers that say it is a false assumption and a myth. There are numbers and stats to be found on each side of the coin about road diets. It is difficult to compare road diets since they each have their own particular shapes and sizes. A road diet might do well in one jurisdiction and do poorly in another. You cannot usually carte blanche declare a road diet as good or bad, and instead would need to look at the circumstances and situation involved.

Back to my analogy about food diets to the nature of road diets. Are all food (roadway) diets bad? Nope. Are some food (roadway) diets bad? Yes. Is a food (roadway) diet that is good for Joe (city X) necessarily also good for Samantha (city Y)? No. Should we be willing to consider a food (roadway) diet? Yes. Is one food (roadway) diet the same as another? Not usually. And so on.

What does this have to do with AI self-driving cars?

At the Cybernetic AI Self-Driving Car Institute, we are developing AI software for self-driving cars.

One aspect that is considered an "edge" problem involves the nature of road diets as it relates to AI self-driving cars. It is considered an edge problem since it is not at the core of what most of the auto makers and tech firms are focusing on. They pretty much are focused on the rudiments of getting the AI to drive a self-driving car. The use of an AI self-driving car in a road dieted situation is not at the core of the driving task, in their view, and can be dealt with at a later time (it is considered on the corner or edge of the core problem being solved).

I'd like to first clarify and introduce the notion that there are varying levels of AI self-driving cars. The topmost level is considered Level 5. A Level 5 self-driving car is one that is being driven by the AI and there is no human driver involved. For the design of Level 5 self-driving cars, the auto makers are even removing the gas pedal, brake pedal, and steering wheel, since those are contraptions used by human drivers. The Level 5 self-driving car is not being driven by a human and nor is there an expectation that a human driver will be present in the self-driving car. It's all on the shoulders of the AI to drive the car.

For self-driving cars less than a Level 5, there must be a human driver present in the car. The human driver is currently considered the responsible party for the acts of the car. The AI and the human driver are co-sharing the driving task. In spite of this co-sharing, the human is supposed to remain fully immersed into the driving task and be ready at all times to perform the driving task. I've repeatedly warned about the dangers of this co-sharing arrangement and predicted it will produce many untoward results.

Let's focus herein on the true Level 5 self-driving car. Much of the comments apply to the less than Level 5 self-driving cars too, but the fully autonomous AI self-driving car will receive the most attention in this discussion.

Here's the usual steps involved in the AI driving task:
- Sensor data collection and interpretation
- Sensor fusion
- Virtual world model updating
- AI action planning
- Car controls command issuance

Another key aspect of AI self-driving cars is that they will be driving on our roadways in the midst of human driven cars too. There are some pundits of AI self-driving cars that continually refer to a utopian world in which there are only AI self-driving cars on the public roads. Currently there are about 250+ million conventional cars in the

United States alone, and those cars are not going to magically disappear or become true Level 5 AI self-driving cars overnight.

Indeed, the use of human driven cars will last for many years, likely many decades, and the advent of AI self-driving cars will occur while there are still human driven cars on the roads. This is a crucial point since this means that the AI of self-driving cars needs to be able to contend with not just other AI self-driving cars, but also contend with human driven cars. It is easy to envision a simplistic and rather unrealistic world in which all AI self-driving cars are politely interacting with each other and being civil about roadway interactions. That's not what is going to be happening for the foreseeable future. AI self-driving cars and human driven cars will need to be able to cope with each other.

Returning to the topic of road diets, let's consider how road diets are related to AI self-driving cars.

First, some pundits argue that we ought to consider implementing road diets as a means to support the advent of AI self-driving cars.

It is assumed that we will gradually have lots and lots of AI self-driving cars on the roadways. Furthermore, those AI self-driving cars are likely to be put to use on a non-stop 24x7 basis. You'll see AI self-driving cars cruising back-and-forth, providing all kinds of ridesharing services for us. You might use an AI self-driving car to take your kids to school and pick them up after classes to drive them home (you won't need to do the driving, instead just send the AI self-driving car). Or, maybe send your AI self-driving car to pick-up that pizza for dinner. Etc.

Some believe that we should consider blocking off many downtown areas to reduce the amount of car traffic and increase the amount of foot traffic and biking that can occur. The use of a road diet approach would presumably aid in this notion.

You might then have possibly well-coordinated AI self-driving cars that are streaming back-and-forth throughout these road dieted locations. Those AI self-driving cars are picking up people or goods, and delivering people or goods. They are well-coordinated in that perhaps the use of V2V (vehicle-to-vehicle communications) and V2I (vehicle-to-infrastructure) communications has allowed them to ascertain which ones are going on which roads, and otherwise align their efforts.

This seems sound in theory and nearly utopian.

As mentioned though, there will be a long-time overlap of human driven cars and AI self-driving cars. Would both human driven cars and AI self-driving cars be both allowed into these road dieted locations? If so, the human driven cars would presumably have a more difficult time coordinating their driving activities than would the only-AI driven cars. Also, the variability of how the driving of the human driven cars would occur in the dieted locations is likely higher than the AI self-driving cars (in essence, it would be possible to have the AI self-driving cars driving in a special "road diet mode" while in a road dieted location).

It is likely that these road dieting efforts will encounter many of the same qualms already expressed about today's road diets. The impact might be that the road congestion generated becomes untenable. It could be that the traffic delays generated become untenable. And so on.

You've ordered a pizza delivery to your downtown apartment which is nestled in a road dieted location. Turns out that the AI self-driving car trying to reach you has been delayed in the road diet area due to the volume of cars trying to push through that roadway. You are upset about the delay in getting your pizza!

I realize that the delay in getting a pizza is somewhat silly perhaps. I just used the pizza example to illustrate what might take place. You can substitute the pizza delivery with let's say medicine being delivered to an elderly person living within the road dieted zone. That obviously ups the stakes in understanding the impact that traffic delays might cause.

Some of the disadvantages of road diets can be likely better controlled via the use of AI self-driving cars. For example, the spillover effect can be potentially reduced by informing the AI systems of self-driving cars that they are not supposed to try and avoid the road diet by taking side streets. This could be possibly pumped to the AI via the on-board OTA (Over-the-Air) electronic communications, which usually involve providing updates or patches to the self-driving car.

In general, the aspects of adjusting downtown areas onto a road diet has both its plusses and minuses. If you could ban human driven cars from those road dieted areas, it might make for a more orderly use of the available car lanes. Whether humans will put up with being banned from driving in those areas would seem like an open question and one that might generate a lot of public debate and controversy.

Even if you could ban human drivers from those areas, you still need to consider how much car traffic you are anticipating. I say that because even if you have only AI self-driving cars allowed into a road dieted location, this does not somehow magically overcome the volume and timing of the car traffic axiomatically. You can only get so much water to flow through a pipe of a certain size. The same would be true about the number of car lanes available and the volume of the AI self-driving cars that are being sought to get into and out of the road dieted area at any point in time.

The other major aspect to consider about AI self-driving cars and road diets consists of the specialized driving nature of traversing and using a road dieted location.

There are some AI developers that say there is nothing unusual or new about driving in a road dieted location. In their book, if an AI self-driving car can navigate a normal road, it should be able to do the same when navigating a road dieted location. I consider this to be a head-in-the-sand belief and one that can bode for problems when AI self-driving cars get themselves into such specialized circumstances.

We believe that there is more to dealing with a road dieted location than just an everyday driving routine. Road dieted locations do have a specialized element and therefore merit specialized AI capabilities to properly undertake.

As already mentioned, one aspect would be the driving practices of the AI self-driving car. In a savvy AI system, if there is a roadway bottleneck, the AI will try to find a means to get around the bottleneck. But, in the case of a road diet, it might be that by-design the self-driving cars are being asked to refrain from trying to spill over into nearby neighborhoods. The AI would need to be able to get notified of this driving condition and be able to adjust accordingly.

Another aspect involves whether the road dieted location might have special cut-out areas that are intended for loading and unloading. It is expected that for the safe and efficient delivery of goods and people, there will be various street cut-out areas to allow for loading and unloading, more so than typically is available today. The advent of large volumes of deliveries via AI self-driving cars and ridesharing will increase the need for these specialized zones.

The AI self-driving car needs to be versed in approaching, stopping, and then resuming a car driving journey in these road dieted locations. This must be done with the upmost safety and with the realization that there will likely be a significant presence of both pedestrians and bicyclists.

The AI self-driving car also needs to be ready to cope with the V2V and V2I electronic communications that are likely to be occurring in those road dieted locations, sifting through what might be a voluminous amount of information and coordination aspects.

Another aspect involves the potential for pranking of an AI self-driving car.

It is anticipated that humans might try to "prank" AI self-driving cars, doing so by for example stepping in front of an AI self-driving car to get it to come to a sudden stop. This might be done just for fun or sport, and not due to genuinely needing to get the AI self-driving car to come to a halt. The odds are that this kind of pranking will occur even more so in a road dieted location, due to the higher volume of nearby pedestrians and bike riders. The AI system needs to be versed in dealing with the pranks.

If there is human driven car traffic allowed into the road dieted area, the AI needs to be prepared to contend with the variabilities of what those human drivers might do. A savvy AI system needs to be versed in defensive driving techniques overall. Within a road dieted location, there are various specific defensive aspects that the AI should be further have available in its driving capabilities.

Another potential difficulty for an AI self-driving car would be the encountering of a road dieted location for the first time. If the AI self-driving car did not realize beforehand there was a road dieted location on its driving journey, perhaps it is not marked as such on a map or GPS, the AI system needs to detect that a road dieted location exists and that the self-driving car has entered into it. Once having discerned and mapped out the road dieted location, it could potentially add this aspect to its repertoire as part of the Machine Learning capabilities.

Road diets, they are a coming. The emergence of AI self-driving cars will likely promote the adoption of lane reductions and road diets. This should not be done blindly. A road diet can be a boon to a local area or become a nightmare. Either way, if a road diet is instituted, the AI of the self-driving car needs to be ready to cope with the particulars of a road dieted location. It is important to make sure that the AI is beefed-up and not too slim on how to best and safely drive in an area that's gotten a slenderized road diet.

CHAPTER 10
WRONG WAY DRIVING
AND
AI SELF-DRIVING CARS

CHAPTER 10

WRONG WAY DRIVING
AND
AI SELF-DRIVING CARS

I sheepishly admit that I have been a wrong way driver. There have been occasions whereby I drove the wrong way, doing so luckily without leading to any undesirable outcome, and for which I certainly regret having mistakenly gone astray. It has happened on several instances inside parking garages or parking lots. I'd dare say that many have made the same error and were likely as confounded by poor signage and convoluted paths as I had been. Fortunately, I didn't go up a down alley and nor did I go down an up alley. I just ended-up going against the alignment of cars in parking spots and quickly realized that I must be going in the wrong direction.

When you suddenly realize that you are heading in the wrong direction, it can be relatively disorienting. How did I get mixed up? Did I miss seeing a sign that warned about going in this direction? The next thought that you have is what to do about the situation. Should you continue forward, even though there is now a chance that you'll come head-to-head with a car that is going in the correct direction? Or, should you back-up, which at least then has your then car going in the proper direction, but a lengthy effort of backing up can have its own dangers.

You can also potentially stop the car wherever you happen to be. At least a stopped car would hopefully be less chancy of sparking a car accident than one that is in-motion and going the wrong way. I'm not suggesting that being stopped is necessarily a safe idea and it could still put you and other cars in danger. Even if you do come to a stop, you obviously cannot just sit there until the cows come home and will ultimately need to decide what to do about the situation.

For most people, I'd bet that they usually are quick to consider turning around. In essence, as soon as practical, try to get your car turned around and headed toward the proper direction. You might do so by coming to a stop first, and then progressively try to make a U-turn by going back-and-forth, assuming that the space in which turning around is tight. If there is abundant room to turnaround, the matter of doing so becomes quite simplified and involves making a U-arch in as swift a movement as you can.

It always seems that just as you start to turnaround the car, another car will come toward you. They then wait for you to make your turnaround. You can usually feel the eyes of the other driver boring at you as you "waste" their time while turning around. The other driver probably thinks that you are quite a clod to have gotten yourself into such a predicament. I even had one driver that honked their horn at me when I was in one instance of turning around – I failed to understand the value of honking the horn since I obviously already knew that I was going in the wrong direction and was trying to rectify the circumstance. Maybe the driver was honking their horn in appreciation for my valiant efforts of turning around (I realize that is the glass-is-half-full perspective of the universe).

Fortunately, I've not personally gone the wrong way on a freeway, nor on a highway or a regular street.

I've certainly known of such wrong way instances that were committed by others. Just a month ago, a wrong way driver at 2:00 a.m. got onto two of the major freeways here in Southern California, the I-5 and the I-110, and proceeded to drive at speeds of 60 to 70 miles per hour. The crazy driver side swiped some other vehicles during the ordeal. The police were brave and actually chased after the driver. It is one thing to be a police officer chasing a speeding car that is going in the proper direction, which already includes a lot of danger, but imagine the heightened risks of chasing after a driver that is going the wrong way and at high speeds. The late hour was fortunate since there wasn't much traffic on the freeways and the driver ultimately was caught (they were DUI, plus driving a stolen car).

I've confronted situations involving a wrong way driver coming at me. One of the most scary and vivid such memories involved a vacation trip to Hawaii with my family. We had rented a car on Maui and were driving around to see the beauty of the island. Going along one of the major highways, Haleakala Highway, there was a grass median that separated the westward side from the eastward side of the road. The grass median was banked and the other side of the road was several feet higher, allowing therefore for seeming protective coverage from anyone veering into the other side. There wasn't any fence or structural barrier dividing the two directions.

The kids were having a great time in the back of the car and relished our being in Hawaii. As I attentively watched the road up ahead, all of a sudden, I saw a car from the upper banked roadway that erratically veered across the grassy median and was now entering into my stretch of road, coming straight at me, barreling along at around 50 to 60 miles per hour toward me. Since I was going the same rate of speed, we were quite rapidly approaching each other, completely going head-to-head.

This is one game of chicken you never want to be involved in.

It was one of those moments in life where time seems to nearly standstill. It was happening so fast that I wasn't even mentally able to digest it fully. My instincts were to try and avoid the car by myself veering onto the grassy median, figuring maybe that was the safest place to be. I could have veered to my right into the slow lane of the highway, but I thought I'd still be a target of the wayward driver. I guessed that maybe the nutty driver might opt to switch into the other lane, perhaps desperately trying to avoid the head-on collision of our cars, and so the grassy median might have been clear. I doubted that we would both meet in the grassy median and was guessing that the other driver would stay on the highway, even if going in the wrong direction.

Just as I was about to make a "panic" swerve up onto the grassy median, in a split second of amazement, I observed the other driver doing the same. I decided to therefore stay in my lane and veer toward my slow lane, aiming to provide as much space between me and the other driver. Sure enough, we zipped past each other with just a few feet to spare. He was on the grassy median and then proceeded further upward and returned to his proper lanes.

The whole matter transpired in a few seconds and I almost doubted my own sanity that it even happened at all. There wasn't any other traffic nearby and so there weren't any other third-party witnesses. The other driver had utterly threatened my life and the lives of my family. Meanwhile, the kids in the backseat were oblivious to the ordeal and had kept laughing and singing throughout those highly tense brow sweating moments.

I'll never know what was in the mind of the other driver. Why did they come down onto my stretch of the highway? What made them opt to go back onto the grassy median, rather than somehow trying to stay going on the highway in the wrong direction? Did this all happen by "accident" in that the driver somehow just messed-up, or was this some kind of intentional act for "fun" or "sport" that the driver had in mind? It only took a few seconds for the entire sequence to reveal itself, and yet to this day I remember it as though it took hours to occur and forever will be one of the scariest driving moments of my life.

There was one other notable wrong way incident that I was involved in and my luck held true that nothing untoward happened. This one is rather incredible and beyond the norm.

While working on my Ph.D., I was doing research on the cognitive ability of air traffic controllers as part of a grant focusing on the Human Computer Interface (HCI). The questions being explored involved how the air traffic controllers made use of their radar systems for tracking air traffic. How much did the air traffic controller need to keep in their mind? To what degree did the radar scopes aid or hinder their ability to route air traffic? What kinds of improvements could be made in the radar systems and the interface so that it would enhance the abilities of the air traffic controllers?

I had at first had air traffic controllers come to our research lab at the university and take various cognitive tests. It was impressive how much of a 3D mental model they could create in their minds, unaided by any system at all. I would tell an air traffic controller that a plane was entering into their air space at such-and-such speed and going in such-and-such direction at such-and-such height. I would continue to add more such flights into the airspace, all imaginary, and wanted to see how many such flights they could mentally handle. The twist too was that the air traffic controller was to imagine where the planes are as time ticks along. It is now say five seconds since those planes each entered into your air space, and I'd ask them where each plane was and whether there was any danger of planes colliding.

Eventually, I realized that it would be advantageous to go observe the air traffic controllers in action. I got permission to go watch the air traffic controllers at LAX (Los Angeles International Airport), considered one of the busiest airports in the United States. These air traffic controllers were considered the top echelon of air traffic controllers, often having worked their way up from other much smaller airports that had much less air traffic and complexity.

I wanted to contrast the top air traffic controllers with those that were still working their way up the controller ladder. So, I got permission to visit a relatively small airport and observe the air traffic controllers there. A fellow researcher and I drove out to the airport together. It was a very foggy night and when we arrived at the airport the fog cloaked most of the airport. We arrived at the airport gate and the security guard told us we could drive directly out to the airport tower. He cautioned us to make sure that we obey all traffic signs and drive at a slow speed. This seemed prudent to us and we agreed to do so. My fellow researcher was driving the car at the time.

Well, before I say what happened next, allow me to offer my personal "excuse" about what was taking place so that you won't judge me too harshly. It was so foggy that you could hardly see your hand in front of your face. We drove along at some snail-like speed of maybe 3-5 miles per hour and kept our eyes peeled. We had rolled down the windows of the car in hopes of being better able to see through the fog. The headlights were bouncing their light off the fog particles and we really could not see much of what was ahead of us.

While crawling along, we began to see a colored light embedded in the roadway just a few feet up ahead of us. We could also see some painted lines on the roadway. Turns out, we were driving on a runway!

That's a rather stunning wrong way story, I believe. How many people do you know that have driven their cars onto a runway? When we realized that we were on a runway, you can imagine that the blood drained from our faces and we both looked at each other in shock. The fog was so thick that we hadn't realized we had meandered onto the runway and we also had no idea which direction would get us off the runway. It turns out too that it was considered an "active" runway that planes could take-off or land upon. Fortunately, the thick fog had temporarily closed-off any flights from landing or taking off.

Of course, I'm alive today to tell the story, and we were able to eventually find the road that led to the airport tower. For a few moments though, we had an encounter of a frightful nature and agreed not to tell anyone about it at the time. Our personal code of a "statute of limitations" on speaking of the matter has run its course and so I am able to tell the story now. I chock the whole experience to the braze nous of youth.

One last quick aspect about driving the wrong way. As a society, we seem to have a fascination with wrong way driving. There are numerous movies and TV shows that depict driving the wrong way. It seems that nearly any cop related movie or spy related movie that is a blockbuster has to have its own car chase that involves going the wrong way. One of my favorite such scenes occurred in the movie Ronin, encompassing an elaborately staged and lengthy sequence of going the wrong way on freeways and in tunnels, etc.

In terms of why people drive the wrong way, here's some reasons:

- Drunk driving

- Confused driver

- Inattentive driver

- Shortcut driver

- Thrill-seeker driver

- Etc.

There has been extensive research about how to design off-ramps and on-ramps to try and prevent confused or inattentive drivers from going the wrong way. It can be relatively easy to get confused when driving in an area that you are unfamiliar with and inadvertently go up an off-ramp. Going down an on-ramp is usually a less likely circumstance since the car driver would need to make some sizable contortions to get their car positioned to do so.

Going the wrong way on a one-way street would be another common means of wrong way driving. I knew one fellow student in college that used to take a one-way street the wrong way in order to get to campus faster.

He loudly complained that the right-way was more convoluted and added at least ten to fifteen minutes to his driving commute. According to him, the one-way was rarely used by other drivers and so he felt comfortable going the wrong way on it. In this case, he was convinced that there was nothing wrong with his shortcut and the "problem" was that the city improperly allocated the street as a one-way in the wrong direction.

As far as I know, he lucked out and never got into a car accident on that one-way. He was proud of the fact that he drove that wrong way for several years and never once got a ticket (well, he never got caught).

The point being that there are some cases whereby a driver goes the wrong way by intent.

My fellow student did so as a shortcut, though I always suspected that maybe he was a bit of a thrill-seeker and got a kick out of going the wrong way. His efforts were completely illegal. He endangered not only himself, but anyone else that was in his car during his trickery and could have endangered any cars that were driving the right-way on that one-way street.

When I was with my family in Hawaii, we had another "wrong way" circumstance arise, though it was thankfully much less eventful than my head-to-head situation.

We were heading up to a remote waterfall and we had to take a winding road that made its way through a thick jungle. I had rented a jeep, just in case the road became difficult to drive on. There was one road that was a one-way up to the waterfall, and a second road that was a one-way down from the waterfall (each being one lane only).

The rental agent handed me the keys to the jeep and then offered a word of advice. She told me that portions of the winding road were washed out by recent storms. As such, there would be areas that I would have to drive on the other road, the one that went in the opposite direction. I was a bit dismayed at this bit of news. I clarified that she was telling me to illegally drive, doing so by going the wrong way. She shrugged it off and said that everyone knew about it and it was usable and practical advice.

There are a mixture of circumstances involving drivers that go the wrong way by mistake while other situations involving a driver that intentionally goes the wrong way. Those that are intentionally going the wrong way might do so under-the-table and without any authority to do so, while in other instances it is conceivable that a driver might be purposely instructed to go the wrong way.

According to statistics by the NHTSA (National Highway Traffic Safety Administration), there are about 350 or so deaths per year in the United States due to wrong way driving. Any such number of deaths is regrettable, though admittedly it is a relatively smaller number of deaths than by other kinds of driving mistakes (there are about 35,000 car related deaths per year in the U.S.). There doesn't seem to be any reliable numbers about how many wrong way instances there are per year and such instances are usually unreported unless there is a death involved.

The total number of miles driven in the United States is estimated at around 3 trillion miles per year. One would guess that driving the wrong way happens daily and amounts to perhaps some notable percentage of that enormous number of driving miles.

Fortunately, it would seem that the number of actual accidents due to wrong way driving is quite small, but this is likely due to the wrong way driver quickly getting themselves out of their predicament and also the reaction of right-way drivers to help avoid a collision. In essence, it might not be happenstance that the wrong way driving doesn't produce more problems. It seems more likely that it is due to human behavior of trying to overt problems when a wrong way instance occurs.

What does this have to do with AI self-driving cars?

At the Cybernetic AI Self-Driving Car Institute, we are developing AI software for self-driving cars. There are two key aspects to be considered related to the wrong way driving matter, namely how to avoid having the AI self-driving car go the wrong way, and secondly what to do if the AI self-driving car encounters a wrong way driver.

Allow me to elaborate.

I'd first like to clarify and introduce the notion that there are varying levels of AI self-driving cars. The topmost level is considered Level 5. A Level 5 self-driving car is one that is being driven by the AI and there is no human driver involved. For the design of Level 5 self-driving cars, the auto makers are even removing the gas pedal, brake pedal, and steering wheel, since those are contraptions used by human drivers. The Level 5 self-driving car is not being driven by a human and nor is there an expectation that a human driver will be present in the self-driving car. It's all on the shoulders of the AI to drive the car.

For self-driving cars less than a Level 5, there must be a human driver present in the car. The human driver is currently considered the responsible party for the acts of the car. The AI and the human driver are co-sharing the driving task. In spite of this co-sharing, the human is supposed to remain fully immersed into the driving task and be ready at all times to perform the driving task. I've repeatedly warned about the dangers of this co-sharing arrangement and predicted it will produce many untoward results.

Let's focus herein on the true Level 5 self-driving car. Much of the comments apply to the less than Level 5 self-driving cars too, but the fully autonomous AI self-driving car will receive the most attention in this discussion.

Here's the usual steps involved in the AI driving task:
- Sensor data collection and interpretation
- Sensor fusion
- Virtual world model updating
- AI action planning
- Car controls command issuance

Another key aspect of AI self-driving cars is that they will be driving on our roadways in the midst of human driven cars too. There are some pundits of AI self-driving cars that continually refer to a utopian world in which there are only AI self-driving cars on the public roads. Currently there are about 250+ million conventional cars in the United States alone, and those cars are not going to magically disappear or become true Level 5 AI self-driving cars overnight.

Indeed, the use of human driven cars will last for many years, likely many decades, and the advent of AI self-driving cars will occur while there are still human driven cars on the roads. This is a crucial point since this means that the AI of self-driving cars needs to be able to contend with not just other AI self-driving cars, but also contend with human driven cars. It is easy to envision a simplistic and rather unrealistic world in which all AI self-driving cars are politely interacting with each other and being civil about roadway interactions. That's not what is going to be happening for the foreseeable future. AI self-driving cars and human driven cars will need to be able to cope with each other.

Returning to the topic of driving the wrong way, let's first consider the possibility of an AI self-driving car that happens to go the wrong way.

Some pundits insist that there will never be the case of an AI self-driving car that goes the wrong way. These pundits seem to think that an AI self-driving car is some kind of perfection machine that will never make any mistakes. I suppose in some kind of utopian world this might be the case, or perhaps for a TV or movie plot it might the case, but in the real-world there are going to be mistakes made by AI self-driving cars.

You might be shocked to think that an AI self-driving car could somehow go the wrong way. How could this happen, you might be asking. It seems incredible perhaps to imagine that it could happen.

The reality is that it could readily happen.

Suppose there is an AI self-driving car, dutifully using its sensors, and is scanning for street signs, but fails to detect a street sign that indicates the path ahead is considered a wrong way direction. There are lots of reasons this could occur.

Maybe the street sign is not there at all and it has fallen down, or vandals had taken it down a while ago.

Or, it might be that the street sign is obscured by a tree branch or maybe it is so banged up and has graffiti that the AI system cannot recognize what the sign is.

Maybe the sign can be only partially seen and does not present itself sufficiently to get a match to the Machine Learning (ML) that was used to be able to spot such signs.

Perhaps the weather conditions are such that it is heavily raining, and the sign cannot be detected or perhaps it is snowing and there is a layer of snow obscuring the signs. And so on.

I assure you, there are lots of plausible and probable reasons that the AI might not detect a street sign that warns that the self-driving car is about to head the wrong way.

You might be thinking that it doesn't matter if the AI is able to detect a sign, since it would certainly have a GPS and map of the area and would realize that the road ahead is one that would involve going the wrong way.

Though it is certainly handy for the AI to have a map of an area and a GPS capability, you cannot assume that a map will always be available and also that the GPS won't necessarily have anything to do about warning of a wrong way up ahead. Currently, the focus for the auto makers and tech firms involves developing elaborated maps of localized areas and then having their trials of the AI self-driving cars take place in a geofenced area.

Once we have widespread AI self-driving cars, I don't think we should be basing their emergence on having mapped every square inch of the world in which they are driving. There are many that are trying to do so, but I am saying that a true Level 5 self-driving car should not be dependent upon having a prior map of wherever it is going. I assert that humans drive in places whereby the human driver has no map at all beforehand, and yet they are still able to sufficiently drive a car. That's the target of a Level 5, in my opinion, namely being able to drive a car in the manner that a human can drive a car.

In short, I am claiming that there are going to be circumstances in which an AI self-driving car is going to end-up going the wrong way. This would happen due to the AI not being able to discern the roadway situation and not having a prior map that would otherwise forewarn that a wrong way is up ahead.

You might still fight me about this notion, but I'll add another twist to see if I can convince you of the possibility of an AI self-driving car getting caught up going the wrong way. Remember earlier that I mentioned I have gone the wrong way in various parking structures and parking lots? I'd be willing to bet that the same kind of wrong way heading could happen to an AI self-driving car.

I doubt that parking structures and parking lots will be mapped to the degree that our freeways, streets, and highways are. As such, the AI self-driving car when encountering a parking lot or parking structure, might well end-up failing to spot signs about the proper direction and could get itself mired in going the wrong way.

A techie might respond by saying that the parking structure or parking lot opt to have some form of electronic communication that would provide directions to the AI self-driving car. I agree that we might well see such electronics being added into all kinds of structures or buildings into which an AI self-driving car might be able to drive. But, I wouldn't bet on it always being available, and furthermore even if it happens the odds are that it will take place slowly over time, and meanwhile there will be structures that do not have such a communications setup.

I'll offer one other comment about this notion of going the wrong way. Are you willing to bet that there will never be a situation involving an AI self-driving car that finds itself going the wrong way? I ask because if the AI self-driving car is not ready to tackle such a predicament, and it is because you are so sure that it will never happen, well, I'd not like to be in or near that AI self-driving car that has gotten itself into such a fix and then is unaware of it or does not know what to do about it.

The auto makers and tech firms are so busy trying to get AI self-driving cars to simply drive the right way on roads, they generally have considered this aspect of dealing with driving the wrong way to be an edge problem. An edge problem is one that is not considered at the core of what you trying to solve. We're not quite so convinced that this should considered an edge per se and that it might well happen more than you might think.

A proficient AI self-driving car needs to be able to detect that it has gotten itself into a wrong way driving situation.

The detection can potentially occur by the sensory input and interpretations of the AI. Once you are immersed in a wrong way driving situation, there are often telltale clues that something is amiss. As mentioned earlier in my story, the wrong way in a parking lot can be potentially detected by realizing that the parked cars are parked in an orientation away from your direction of travel. Another might be that there are other roadway signs for which the AI self-driving car is only seeing the backside of the sign, or other signs that have arrows that point in a direction other than the direction of the AI self-driving car.

The AI might also detect cars that are coming straight toward the self-driving car, similar to my example earlier of the game of chicken that I had with a wrong way driver. There might be other surroundings related aspects such as the movement of pedestrians, the motion of bicyclists, and other environmental aspects that can be used to detect a wrong way situation.

The sensor fusion is crucial at this juncture since it is often difficult to ascertain via one indicator alone that the AI self-driving car is going the wrong way. It might be a multitude of indications coming from a multitude of the sensors, all of which needs to be combined and considered during the sensor fusion portion of the AI system driving task.

It could be too that the AI self-driving car might get alerted via electronic communications. There could be other AI self-driving cars nearby and those self-driving cars might have detected that your AI self-driving car is going the wrong way. They could potentially communicate via V2V (vehicle-to-vehicle communication) and let the AI of your self-driving car know that it is heading in the wrong direction. There might also be V2I (vehicle-to-infrastructure) that could also be alerting the AI of the self-driving car.

If the AI somehow becomes aware of the matter, it then needs to update the virtual world model and prepare an action plan of what to do. This is where the AI might then opt to do the same actions that a human driver might do in such a situation, including coming to a stop, or perhaps moving ahead slowly, or maybe trying to execute a U-turn, etc. The AI needs to determine what is the prudent and safe approach to get itself out of the predicament.

One other consideration in this matter involves the role of a human occupant that might be inside the AI self-driving car. So far, we've assumed that the AI is doing the driving task and doing so without any interaction with humans. I've predicted that the AI of self-driving cars will be interacting extensively with human occupants, doing so for conversational purposes and at times for aspects related to the driving.

I am not suggesting that the human occupants will be guiding the AI as to the driving task. That's not what should be happening in a true Level 5 self-driving car. It is up to the AI to drive the car. But, this does not mean that the AI cannot interact with the human occupants and thus perhaps alter or shape the driving based on that interaction. If you were being driven a human chauffer, you would likely interact with the person, and yet the chauffer is still the driver and has direct and sole access to the driving controls.

It could be that a human occupant might notice that the AI self-driving car has gone the wrong way. In which case, what should the human occupant do? Presumably, the human occupant could engage the AI in a dialogue and indicate that the AI has gone the wrong way. This would likely be an urgent discussion. The AI cannot though blindly assume that the human occupant is correct, in the same sense that a human chauffer would not necessarily blindly believe or abide by whatever a passenger in the car might say.

I'd like to provide an additional variation on the rational for the wrong way driving of an AI self-driving car. There might be situations wherein the AI self-driving car is purposely supposed to go the wrong way. Remember my personal example that I mentioned about driving a jeep in Hawaii to get up to a waterfall? In that case, I was told by an authority figure that there might be portions of the road that would involve my having to go the wrong way.

There have been other situations involving my being told to drive the wrong way. A car accident had blocked part of a major coast highway and the police were directing traffic to go the wrong way on a diversion street. They were forcing traffic to go up a one-way road in the wrong direction. I admit a bit of hesitation when I abided by the police officer's instruction, but I figured the police knew what they were doing. In this case, the police had made sure that this was a safe path to take.

I mention this aspect because suppose that an AI developer has decided that an AI self-driving car should never go the wrong way. This might be done under the naïve belief that since it is dangerous and wrong for an AI self-driving car to go the wrong way that it should be restricted from ever doing so. There are going to be circumstances that involve an AI self-driving car driving "illegally" by doing something such as going the wrong way. This is yet another reason why the AI needs to be prepared to do so.

Now that we've covered the aspects of a wrong way driving AI self-driving car, let's shift our attention to the situation of an AI self-driving car that is confronted by a wrong way driver.

I think you can probably agree with me that there is a likely chance that an AI self-driving car might ultimately encounter a wrong way driver. As I've mentioned, there is going to be a mix of human drivers and AI self-driving cars, occurring for many years. The odds of a human driver going the wrong way towards an AI self-driving car seems reasonably likely. It could happen because the human driver has made a mistake, or it could be that the human driver is drunk or DUI, or maybe the human driver is trying to get away from the police, and so on.

What should the AI do?

It needs to first detect that the wrong way car is headed towards it. Once this detection has occurred, the virtual world model needs to get updated. The AI action planner then can consider various scenarios of how the wrong way situation might play out. This is similar to my harrowing story of being in Hawaii and facing a wrong way driver.

The AI system might decide that it is best to continue forward "as is" or it might decide to take an evasive action. This all depends upon the situation at hand. Is there other nearby traffic? How soon will a collision happen? Which approaches seem to offer the best chances for survival? Etc.

The AI might also be able to confer with other nearby AI self-driving cars. Again, via V2V, it could be that the AI of your self-driving car might either become aware of the wrong way driver by being warned by some other AI self-driving car, or it could be that several AI self-driving cars might band together, momentarily, in an effort to deal with the wrong way driver situation.

There are also the ethical aspects involved in the matter of the AI trying to determine what action to take.

As per the famous Trolley problem, an AI system in such a situation might need to make a "decision" that involves trying to minimize loss of life, and yet it is somewhat ambiguous as to how the AI is supposed to do so. Should the AI opt to swerve into the next lane to avoid the head-on collision with the wrong way driver, but it could be that by swerving into the next lane that the AI self-driving car will collide with another car that is heading in the correct direction. These other innocents in that car might get killed, due to the wrong way driver and due to the choices made by the AI about contending with the wrong way driver.

It might be that any action taken by the AI, even taking no particular action, might end-up with an unavoidable crash. What is the basis for making such a decision?

Some final thoughts about the wrong way driver situations.

Suppose we do eventually have only AI self-driving cars and there are no human driven cars. What then? Well, I still contend that there is a possibility of having an AI self-driving car that can end-up going the wrong way. Thus, the AI systems of self-driving cars should have a provision for dealing with such situations.

I would guess though that by the time we would have all and only AI self-driving cars on our roadways, the odds are that we'd have lots of IoT (Internet of Things) and quite sophisticated V2V, V2I, and even V2P (vehicle-to-pedestrian) electronic communications. As such, the odds of an AI self-driving car going the wrong way would be substantially reduced.

Furthermore, the AI self-driving cars could likely by then coordinate sufficiently with each other in a manner that a wrong way AI car poses no particular concern per se. In essence, the AI of the wrong way driving self-driving car would coordinate with the other AI self-driving cars and in a somewhat seamless fashion get itself out of the predicament. The other AI self-driving cars could actively help to rectify the matter, perhaps slowing down to allow time for the wrong way AI to get itself righted or taking other proactive actions to assist.

I'll end with this somewhat mind-bending thought. In the movies and TV there are those car chases whereby the spy or crook goes the wrong way, and miraculously lives, doing so by magically avoiding the oncoming cars. This is something unlikely to be realistic in today's world of human drivers. If you drove on the wrong way of a busy freeway or highway, I'd dare say that someone is going to get hurt.

In a world of only AI self-driving cars, I suppose you could say that it would be feasible to go the wrong way. Assuming that you have generally perfect V2V and the AI of the self-driving cars are all working in concert with each other, you could in theory purposely go the wrong way, even on a busy road, and do so without incurring any collisions.

Some might even say that this might be a means to more efficiently use our roadways. You could allow AI self-driving cars to use the same roads for both directions and let them figure out how to make it happen. The Golden Gate Bridge adjusts some of the lanes during the peak traffic times to allow for traffic to go one direction and then later in the day shift to the other direction. Nonetheless, there is still only one direction allowed at a time. Imagine a situation whereby we allowed self-driving cars to go in any direction on our roads and go straight head-to-head with other self-driving cars.

Even if this seems theoretically possible, I'd suggest that if you were a passenger in such a self-driving car, you'd have a tough time watching this occur. Then again, will we be so accustomed to believing in the AI of the self-driving cars by then that we would also readily accept them playing this game of chicken? Hard to imagine. For the foreseeable future, wrong way will be wrong way, and wrong way won't be right way. That's my prediction.

CHAPTER 11

WORLD SAFETY SUMMIT

AND

AI SELF-DRIVING CARS

CHAPTER 11
WORLD SAFETY SUMMIT
AND
AI SELF-DRIVING CARS

I was chatting with Jamie Hyneman, notable co-host of the former MythBusters series, during the World Safety Summit on Autonomous Technology (he was the moderator for the event undertaken at the Levi's Stadium in Silicon Valley on October 18, 2018), and we both marveled at the notion that in today's world we all drive around in these metal cans called cars that come within inches of each other at speeds of 80 miles per hour or more, and yet somehow this happens without continuous catastrophic results.

Of course, there are already significant dreadful outcomes and the number of car crashes and car related deaths is abhorrent, for example there are an approximate 37,000 fatalities that occur in the United States alone each year via car related incidents. Given though the volume of cars and the millions upon millions of miles driven, an estimated 3.22 trillion driving miles per year in the U.S. (per the Federal Highway Administration statistics), it is somewhat remarkable that there aren't even more car related deaths.

By-and-large, most of these conventional cars related adverse outcomes can be traced to the driver of the vehicle. In other words, it's not particularly that the car itself had some mechanical fault that led to the horrid outcome, but instead that the human driver by one means or another was the main contributor to the incident.

One of the key questions to be addressed about the advent of AI self-driving cars involves whether or not the newly emerging Autonomous Vehicles (AV's) will be as safe, or safer than, or perhaps less safe than the use of conventional cars. Obviously, the notion and desire are that AI self-driving cars will be at least as safe as conventional cars, and hopefully much safer.

When I say that an AI self-driving car is safe or safer than conventional cars, I'm not especially referring to whether the drivetrain works better or whether the engine works more auspiciously, those are facets that we all pretty much assume will be at least as safe as conventional cars. Instead, the aspect of safety that we're really referring to consists of how the self-driving car will be driven.

Will the AI be able to drive an AI self-driving car as safely or more so than a human driver?

If the auto makers and tech firms cannot assure the public and the regulators that AI self-driving cars are "safe" then the emergence of these AV's will be likely delayed, and progress substantially stinted. Imagine too if AI self-driving cars are allowed fully into the wild (on our public roadways), and it turns out they get into numerous car crashes and people are injured and killed. It is predictable that a backlash of such magnitude could develop that efforts toward AI self-driving cars could become fully undermined and possibly even mothballed.

Marta Thoma Hall, President of Velodyne LiDAR, noted during the World Safety Summit that if you told someone to go across a bridge that is only partially built, their trepidation to do so would certainly be understandable. Yet, there are some in the AI self-driving car industry that don't seem to get the notion that we are indeed asking people at this time to go across a partially built bridge.

During my numerous conference speaking engagements about AI self-driving cars, I often have fellow AI developers that approach me and question why the lay public doesn't trust AI self-driving cars. I am more so surprised that these AI developers assume that people should have blind-faith in the existing capabilities of AI self-driving cars, more so than being at all surprised that the public has hesitation about these innovations.

The tech industry relishes embracing the "fail fast, fail first" mentality, and yet in the case of AI self-driving cars it is quite unlikely that the public and regulators are going to tolerate substantive failure rates. Failing fast when developing a social media site or when crafting a game app might be sensible, but doing the same for AI self-driving cars, which are real-time systems that encompass life-or-death matters, I don't think most of society will have a stomach for it, in spite of what some would say is worthwhile due to the desired outcomes at the end of the rainbow once AI self-driving cars have been perfected.

Furthermore, with the increasing number of firms that are aiming to produce AI self-driving cars (around 50 such firms are signed-up to test drive their self-driving cars on California roads), it would seem to increase the chances that there might be one bad apple in the bunch, meaning that if an experimenter auto maker or tech firm, once granted the privilege to put their AI self-driving car on public roads, causes headline worthy injuries or deaths by their contraption, all of the rest of the auto makers and tech firms trying to also develop and field AI self-driving cars could get tainted by the same sour outcome. One bad apple can regrettably and absolutely spoil this entire barrel.

It is an imperative that safety and AI self-driving cars go together. They must be joined at the hip. Insufficient safety and this ship will sink. Indeed, per the World Safety Summit comments of Christopher Hart, former National Transportation Safety Board (NTSB) Chairman, recall a famous ship that had been heralded as unsinkable. The AI self-driving car community and industry does not want to get itself into a similar bind of thinking that AI self-driving cars are unsinkable.

At the Cybernetic AI Self-Driving Car Institute, we are developing AI software for self-driving cars and are very attuned to the need for safety alertness in the AI systems of self-driving cars. Safety has to be explicitly built into the AI and be considered part of its core design.

Trying to somehow add-on safety considerations for self-driving cars after-the-fact would be foolhardy and more so generally impracticable.

There are some in the AI self-driving car industry that consider safety to be an edge problem, meaning that it is not at the core of the systems effort for producing AI self-driving cars. The focus of these AI developers is that they first are seeking to get the self-driving car to drive along a road and do so with some modicum of capability. They assume that other cars will pretty much stay out of the way of the AI self-driving car and that the traffic and pedestrians nearby will give the self-driving car wide berth. In that sense, those developers are betting on safety by believing in a constrained traffic environment, but that's not the real-world of driving that we all face each and every day on our open roads.

This also brings up the aspect of asserting that AI self-driving cars will lead us to achieving zero fatalities in car related deaths. As I've indicated many times, though the notion of zero fatalities is certainly laudable as a goal, unfortunately it is not achievable and worse too it also has the potential for setting misleading expectations for the public and regulators about AI self-driving cars.

Why aren't zero fatalities realistic? Let's consider car crashes to be divided into those that are avoidable and those that are unavoidable. For an avoidable car crash, the driver, whether human-based or AI, can potentially maneuver the car in a manner that avoids the car crash and therefore presumably avoids potential human injury or death. For an unavoidable car crash, the driver, again whether human-based or AI, will be unable to avoid the car crash, along with the chances of human injury or death, no matter what the driver might try to do.

Suppose a car is being driven down a street at 45 miles per hour and a pedestrian that was standing at the curb suddenly jumps out into the street, doing so with just a split second to go before impact by the car. The physics of the situation bely the aspect that a car crash and injury or death can be avoided. You might say that the driver of the car should have realized that the pedestrian was going to jump off the curb, but even if you detected the pedestrian it could be that the pedestrian made no overt indication of what they were about to do. My point being that no matter how good a driver the human might be or the AI might be, there are still going to be unavoidable crashes.

So, I am asserting that no matter how good the AI might be, it will still find itself driving a self-driving car that will get into unavoidable car crash situations. Now, we might be able to assume that the number of unavoidable car crashes will be a lot less than the number of car crashes involving fatalities today. In essence, if we could parse out the number of car crashes that were unavoidable out of the total number experienced today, the odds are that it is a low number. Lower but still more than zero.

Another facet to be considered involves the level of autonomous capability of the AI and the self-driving car. I'd like to clarify and introduce the notion that there are varying levels of AI self-driving cars.

The topmost level is considered Level 5. A Level 5 self-driving car is one that is being driven by the AI and there is no human driver involved. For the design of Level 5 self-driving cars, the auto makers are even tending toward removing the gas pedal, brake pedal, and steering wheel, since those are mechanisms used by human drivers. The Level 5 self-driving car is not being driven by a human and nor is there an expectation that a human driver will be present in the self-driving car. It's all on the shoulders of the AI to drive the car (a Level 4 is similar though it has a more constrained driving environment in which it is able to drive).

For self-driving cars less than a Level 4 and Level 5, there must be a human driver present in the car. The human driver is currently considered the responsible party for the acts of the car. The AI and the human driver are co-sharing the driving task. In spite of this co-sharing, the human is supposed to remain fully immersed into the driving task and be ready at all times to perform the driving task. I've repeatedly warned about the dangers of this co-sharing arrangement and predicted it will produce many untoward results.

Another key aspect of AI self-driving cars is that they will be driving on our roadways in the midst of human driven cars too. There are some pundits of AI self-driving cars that continually refer to a Utopian world in which there are only AI self-driving cars on the public roads. Currently there are about 250+ million conventional cars in the United States alone, and those cars are not going to magically disappear or become true Level 5 AI self-driving cars overnight.

Indeed, the use of human driven cars will last for many years, likely many decades, and the advent of AI self-driving cars will occur while there are still human driven cars on the roads. This is a crucial point since this means that the AI of self-driving cars needs to be able to contend with not just other AI self-driving cars, but also contend with human driven cars. It is easy to envision a simplistic and rather unrealistic world in which all AI self-driving cars are politely interacting with each other and being civil about roadway interactions. That's not what is going to be happening for the foreseeable future. AI self-driving cars and human driven cars will need to be able to cope with each other.

When addressing the topic of safety and AI self-driving cars, we need to explicitly consider the level of the self-driving car. Let's for the moment define truly autonomous AI self-driving cars as Level 5 and to some degree Level 4, while the Level 3 and below we'll define as co-shared driving and we will not consider it to be autonomous. I realize that you might argue that there is some autonomous driving aspects at the Level 3, but due to the co-sharing aspects of the driving task at Level 3 let's for the moment lump the Level 3 into the co-shared driving and not put it into the truly autonomous category.

In essence, we'll use two categories for AI self-driving cars, those that are co-shared driving and not truly autonomous, and the other category is those that are truly autonomous.

We can then ask the safety question about each of the two classes or categories. This is an important distinction since the answer will differ between the two circumstances. If we lump together all variants of AI self-driving cars, we would not adequately be able to address the safety question due to the commingling of the two quite different classes.

Thus, we have two questions to address:

- What is the acceptable safety for the AI self-driving car that encompasses co-shared human driving?

- What is the acceptable safety of the truly autonomous AI self-driving car?

The word "safety" is a relative term and one that has a lot of loaded baggage.

I routinely fly on commercial airplanes for work purposes and that take me to destinations all around the country and the world. Would I consider these airplanes to be safe? Yes, I would. A colleague of mine is "afraid" of flying and insists that airplanes are not safe (he avoids flying and often takes a train or ship instead). Who is right? Am I right to say that airplanes are safe, or is he right to claim that they are unsafe?

The typical definition for safety involves the indication that by being "safe" you are not likely to be harmed. Using that kind of a definition, you could assert that today's commercial airplane travel is relatively safe since the chances of injury or death due to a plane accident is low. Apparently, my colleague does not share this viewpoint that the chances of injury or death are low, and instead believes the odds to be "high" or at least high enough that it dampens his enthusiasm in a willingness to fly.

Overall, the point being that safety is as much a perception measure as it is a reality measure. The safety aspect is going to be a chance or probability of some kind of harm, and whether you think that something is safe or not will depend upon how much of a chance or probability you think rises above some otherwise "personal" threshold.

Suppose that the chances of a plane crash are one for each of 1.2 million flights, and that the chances of dying are 1 in 11 million by a plane crash. The odds of getting struck by lightning are higher and so is the chances of getting killed by the flu. My colleague should be walking around worried that he is going to get struck by lightning or that he will contract the flu and die, but he does not seem particularly concerned about either of those potential events.

For AI self-driving cars, we are going to contend with both actual quantifiable measures of their safety, along with the public perception of safety. When considering the matter of "safety" and AI self-driving cars, it includes the perception of what is considered as a safe enough threshold, and will vary by segments of the public, by segments of regulators, and other stakeholders too.

I mention this aspect because some AI pundits keep saying that as long as AI self-driving cars can reduce the number of annual car related deaths, presumably the public will accept and even outright embrace the advent of AI self-driving cars. This though does not seem to take into account the nature of how the perception of safety occurs and is shaped.

Let's imagine that the advent of AI self-driving cars reduces the number of annual car related deaths in the United States from around 37,000 to instead say 27,000 (I'm using this as a placeholder number for purposes of discussion herein and not due to any actual prediction per se). That's a whopping decrease of 10,000 deaths and thus about a 30% or so drop in the total number of annual deaths. You might believe that if AI self-driving cars could produce that kind of lessened number of deaths, it would therefore be heralded as a blessing to all.

I would dare say that the public is not going to be able to necessarily see things in that light. Instead, for each self-driving car related death, there is more than likely going to be a severe hand wringing about why the death occurred and why the advent of AI self-driving cars is not providing its promise of zero fatalities. If people are promised zero fatalities, they therefore tend to think that the decrease in the number of car related deaths would drop from 37,000 to 0, all in one fell swoop.

The path of getting toward a lower fatality rate is going to be a tortuous one and something that public perception cannot readily see as an overall improved rate of lessened deaths over conventional cars. Suppose that we could somehow go from 37,000 annual deaths to 35,000 in year one, down to 30,000 in year two, down to 25,000 in year three, down to 20,000 in year four, down to 15,000 in year five, down to 10,000 in year six, and suppose it ended at 5,000 in year seven and thereafter (thus accounting for the number of unavoidable car crashes).

Could the public and the regulators be content with that kind of progression towards better safety due to the advent of AI self-driving cars? It's a tough pill to swallow.

We also then need to consider my earlier point about the distinction between co-shared driving autonomous cars and those that are truly autonomous cars. For the segment of car crashes that will involve the co-shared driving, we need to consider that presumably some percentage of those will be due to the human driver versus the AI driving aspects. Indeed, it could be that the number of car related deaths could go up, rather than down, if you end-up with a struggle taking place between the human driver and the AI system in circumstances of potential crashes.

Imagine that we had 37,000 annual deaths with conventional cars. Suppose we then infuse into society some amount of co-shared driving autonomous cars, along with some percentage of truly autonomous self-driving cars. We then will have a mixture of conventional cars, combined with co-shared driving cars, along with truly autonomous cars. This mixture is going to presumably impact the number of annual car related deaths.

Let's assume the conventional cars portion of the mix continues at the existing anticipated car related deaths rate (though, this can be argued somewhat since the aspect of the now added mixture of co-shared and truly autonomous might impact the rate).

Let's assume that the truly autonomous self-driving cars are able to achieve a near-zero death rate (this is argumentative given the mixture of the other conventional cars and the co-shared driving cars).

What will be the expected rate of car related deaths in the co-shared driving instances? We don't yet know. Will the addition of the AI reduce the chances of the death rate? Maybe, if you assume that the AI will be able to do a better job at the driving task than the humans in conventional cars. Will the AI perhaps increase the chances of the death rate? Maybe, if you take into account that the human is co-sharing the driving task and that there might be hand-off issues and other confusion that regrettably negates the positive reduction in potential deaths by actually increasing the rate of deaths and car crashes.

Thus, when trying to ascertain what the death rate impacts are due to the advent of AI self-driving cars, it is a complex matter since it involves the mixture of conventional car driving, co-shared driving, and truly autonomous driving. You can bet that any car related deaths due to the co-shared driving and the truly autonomous driving will get magnified tremendously and for some it is seen as a bellwether of what is yet to come. Even one car related death in the truly autonomous category can be perceived as an indicator that AI self-driving cars are more so death-traps rather than saviors, and depending upon what ax one has to grind, the circumstance can be used accordingly.

In the very act of defining safety as it relates to AI self-driving cars, we need to be mindful of a multitude of factors.

There are some that are aiming to use the death outcome as the sole measure for defining safety. Thus, they say that we should measure safety by how many self-driving car related deaths there are per year, similar to the measure used with conventional cars.

I would be willing to bet that the public would perceive self-driving car related injuries to also be an element of the safety of AI self-driving cars. In other words, suppose that AI self-driving cars reduced the number of annual car related deaths, but at the same time the number of human related injuries went up. Perhaps the AI self-driving cars were doing a better job at avoiding direct head-on collisions, but in so doing were sideswiping other cars and pedestrians, thus producing more human injuries than before. Would the public be willing to accept a trade-off of the deaths count for the injuries count?

Another factor involves deaths and injuries involving not just humans but also animals. I know that some think it preposterous to potentially include injuries or deaths of animals, and they bristle that such a count could be somehow compared with the counts involving humans. But, once again, let's consider the public perception. If AI self-driving cars are unable to adequately avoid hitting animals, like say someone's pet dog that wandered into the street as a self-driving car came along, will the public at large be willing to accept that the beloved and innocent pet was injured or killed by an AI self-driving car?

There are also the potential aspects of damages to be encompassed by the safety moniker. Let's separate damages from any kind of injuries or deaths. We'll say that damagers are aspects such as cars that can bashed up or that take out street posts or ram into other roadway infrastructure. If AI self-driving cars are let's say able to avoid some amount of human deaths but in the meantime are creating more damages, perhaps by ramming into objects, what would the public perceive as the overall safety of AI self-driving cars?

We are at a crossroads in the AI self-driving car field of trying to grapple with defining safety. It is a vital measure and one that will ultimately be a determiner of the acceptance of AI self-driving cars. Safety encompasses a multitude of factors and it also is something that is measured on a continuum or spectrum, rather than via a single point. There are some efforts within the automotive field to aid in defining safety aspects of AI self-driving cars, including for example ISO 21448 and ISO 26262, along with the myriad of efforts underway by the SAE.

Echoed repeatedly at the World Safety Summit was the belief that collaboration among the many stakeholders of AI self-driving cars is going to be key to reaching a realistic and usable notion of safety. These stakeholders include the auto makers, the tech firms, the regulators, the media, the safety advocacy groups, the industry associations, the researchers, and so on. I've previously called for a form of "coopetition" among such entities, which will aid in assuring that what collectively is devised will also hopefully be collectively supported. So, yes, absolutely, collaboration is needed and keenly sought.

I was particularly heartened during the World Safety Summit when Alex Epstein, Director of Transportation Safety for the National Safety Council pointed out that we should all think about our son or daughter going onto the roadways when we are trying to wrestle with defining safety and AI self-driving cars.

Those of us immersed in the AI self-driving car realm need to keep our eye on the humanity of what we are developing. AI self-driving cars provide the promise of a tremendous technological advancement and are bound to transform society in incredibly advantageous ways, but with that comes momentous responsibility and a duty to be fixated on safety.

APPENDIX

APPENDIX A
TEACHING WITH THIS MATERIAL

The material in this book can be readily used either as a supplemental to other content for a class, or it can also be used as a core set of textbook material for a specialized class. Classes where this material is most likely used include any classes at the college or university level that want to augment the class by offering thought provoking and educational essays about AI and self-driving cars.

In particular, here are some aspects for class use:

o   Computer Science. Studying AI, autonomous vehicles, etc.

o   Business. Exploring technology and it adoption for business.

o   Sociology. Sociological views on the adoption and advancement of technology.

Specialized classes at the undergraduate and graduate level can also make use of this material.

For each chapter, consider whether you think the chapter provides material relevant to your course topic. There is plenty of opportunity to get the students thinking about the topic and force them to decide whether they agree or disagree with the points offered and positions taken. I would also encourage you to have the students do additional research beyond the chapter material presented (I provide next some suggested assignments they can do).

RESEARCH ASSIGNMENTS ON THESE TOPICS

Your students can find background material on these topics, doing so in various business and technical publications. I list below the top ranked AI related journals. For business publications, I would suggest the usual culprits such as the Harvard Business Review, Forbes, Fortune, WSJ, and the like.

Here are some suggestions of homework or projects that you could assign to students:

a) <u>Assignment for foundational AI research topic</u>: Research and prepare a paper and a presentation on a specific aspect of Deep AI, Machine Learning, ANN, etc. The paper should cite at least 3 reputable sources. Compare and contrast to what has been stated in this book.

b) <u>Assignment for the Self-Driving Car topic</u>: Research and prepare a paper and Self-Driving Cars. Cite at least 3 reputable sources and analyze the characterizations. Compare and contrast to what has been stated in this book.

c) <u>Assignment for a Business topic</u>: Research and prepare a paper and a presentation on businesses and advanced technology. What is hot, and what is not? Cite at least 3 reputable sources. Compare and contrast to the depictions in this book.

d) <u>Assignment to do a Startup:</u> Have the students prepare a paper about how they might startup a business in this realm. They must submit a sound Business Plan for the startup. They could also be asked to present their Business Plan and so should also have a presentation deck to coincide with it.

You can certainly adjust the aforementioned assignments to fit to your particular needs and the class structure. You'll notice that I ask for 3 reputable cited sources for the paper writing based assignments. I usually steer students toward "reputable" publications, since otherwise they will cite some oddball source that has no credentials other than that they happened to write something and post it onto the Internet. You can define "reputable" in whatever way you prefer, for example some faculty think Wikipedia is not reputable while others believe it is reputable and allow students to cite it.

The reason that I usually ask for at least 3 citations is that if the student only does one or two citations they usually settle on whatever they happened to find the fastest. By requiring three citations, it usually seems to force them to look around, explore, and end-up probably finding five or more, and then whittling it down to 3 that they will actually use.

I have not specified the length of their papers, and leave that to you to tell the students what you prefer. For each of those assignments, you could end-up with a short one to two pager, or you could do a dissertation length paper. Base the length on whatever best fits for your class, and the credit amount of the assignment within the context of the other grading metrics you'll be using for the class.

I mention in the assignments that they are to do a paper and prepare a presentation. I usually try to get students to present their work. This is a good practice for what they will do in the business world. Most of the time, they will be required to prepare an analysis and present it. If you don't have the class time or inclination to have the students present, then you can of course cut out the aspect of them putting together a presentation.

If you want to point students toward highly ranked journals in AI, here's a list of the top journals as reported by *various citation counts sources* (this list changes year to year):

o   Communications of the ACM

o   Artificial Intelligence

o   Cognitive Science

o   IEEE Transactions on Pattern Analysis and Machine Intelligence

o   Foundations and Trends in Machine Learning

o   Journal of Memory and Language

o   Cognitive Psychology

o   Neural Networks

o   IEEE Transactions on Neural Networks and Learning Systems

o   IEEE Intelligent Systems

o   Knowledge-based Systems

GUIDE TO USING THE CHAPTERS

For each of the chapters, I provide next some various ways to use the chapter material. You can assign the tasks as individual homework assignments, or the tasks can be used with team projects for the class. You can easily layout a series of assignments, such as indicating that the students are to do item "a" below for say Chapter 1, then "b" for the next chapter of the book, and so on.

a) What is the main point of the chapter and describe in your own words the significance of the topic,

b) Identify at least two aspects in the chapter that you agree with, and support your concurrence by providing at least one other outside researched item as support; make sure to explain your basis for disagreeing with the aspects,

c) Identify at least two aspects in the chapter that you disagree with, and support your disagreement by providing at least one other outside researched item as support; make sure to explain your basis for disagreeing with the aspects,

d) Find an aspect that was not covered in the chapter, doing so by conducting outside research, and then explain how that aspect ties into the chapter and what significance it brings to the topic,

e) Interview a specialist in industry about the topic of the chapter, collect from them their thoughts and opinions, and readdress the chapter by citing your source and how they compared and contrasted to the material,

f) Interview a relevant academic professor or researcher in a college or university about the topic of the chapter, collect from them their thoughts and opinions, and readdress the chapter by citing your source and how they compared and contrasted to the material,

g) Try to update a chapter by finding out the latest on the topic, and ascertain whether the issue or topic has now been solved or whether it is still being addressed, explain what you come up with.

The above are all ways in which you can get the students of your class involved in considering the material of a given chapter. You could mix things up by having one of those above assignments per each week, covering the chapters over the course of the semester or quarter.

As a reminder, here are the chapters of the book and you can select whichever chapters you find most valued for your particular class:

<u>Chapter Title</u>

Companion Book By This Author

Advances in AI and Autonomous Vehicles: Cybernetic Self-Driving Cars

Practical Advances in Artificial Intelligence (AI) and Machine Learning

by

Dr. Lance B. Eliot, MBA, PhD

This title is available via Amazon and other book sellers

Companion Book By This Author

Self-Driving Cars:
"The Mother of All AI Projects"

by Dr. Lance B. Eliot, MBA, PhD

This title is available via Amazon and other book sellers

Companion Book By This Author
Innovation and Thought Leadership on Self-Driving Driverless Cars
by Dr. Lance B. Eliot, MBA, PhD

<u>Chapter Title</u>

1 Sensor Fusion for Self-Driving Cars

2 Street Scene Free Space Detection Self-Driving Cars

3 Self-Awareness for Self-Driving Cars

4 Cartographic Trade-offs for Self-Driving Cars

5 Toll Road Traversal for Self-Driving Cars

6 Predictive Scenario Modeling for Self-Driving Cars

7 Selfishness for Self-Driving Cars

8 Leap Frog Driving for Self-Driving Cars

9 Proprioceptive IMU's for Self-Driving Cars

10 Robojacking of Self-Driving Cars

11 Self-Driving Car Moonshot and Mother of AI Projects

12 Marketing of Self-Driving Cars

13 Are Airplane Autopilots Same as Self-Driving Cars

14 Savvy Self-Driving Car Regulators: Marc Berman

15 Event Data Recorders (EDR) for Self-Driving Cars

16 Looking Behind You for Self-Driving Cars

17 In-Car Voice Commands NLP for Self-Driving Cars

18 When Self-Driving Cars Get Pulled Over by a Cop

19 Brainjacking Neuroprosthetus Self-Driving Cars

This title is available via Amazon and other book sellers

Companion Book By This Author
New Advances in AI Autonomous Driverless Cars Self-Driving Cars
by Dr. Lance B. Eliot, MBA, PhD

<u>Chapter Title</u>

This title is available via Amazon and other book sellers

Companion Book By This Author

Introduction to
Driverless Self-Driving Cars

by Dr. Lance B. Eliot, MBA, PhD

This title is available via Amazon and other book sellers

This title is available via Amazon and other book sellers

Companion Book By This Author
Transformative Artificial Intelligence
Driverless Self-Driving Cars
by Dr. Lance B. Eliot, MBA, PhD

This title is available via Amazon and other book sellers

Companion Book By This Author

Disruptive Artificial Intelligence and Driverless Self-Driving Cars

by Dr. Lance B. Eliot, MBA, PhD

Chapter Title

This title is available via Amazon and other book sellers

Companion Book By This Author

State-of-the-Art
AI Driverless Self-Driving Cars

by Dr. Lance B. Eliot, MBA, PhD

Chapter Title

This title is available via Amazon and other book sellers

Companion Book By This Author

Top Trends in
AI Self-Driving Cars

by Dr. Lance B. Eliot, MBA, PhD

<u>Chapter Title</u>

This title is available via Amazon and other book sellers

This title is available via Amazon and other book sellers

Companion Book By This Author

Crucial Advances for AI Self-Driving Cars

by Dr. Lance B. Eliot, MBA, PhD

This title is available via Amazon and other book sellers

Companion Book By This Author

Sociotechnical Insights and AI Driverless Cars

by Dr. Lance B. Eliot, MBA, PhD

This title is available via Amazon and other book sellers

Companion Book By This Author

Pioneering Advances for AI Driverless Cars

by Dr. Lance B. Eliot, MBA, PhD

This title is available via Amazon and other book sellers

Companion Book By This Author

Leading Edge Trends for AI Driverless Cars

by Dr. Lance B. Eliot, MBA, PhD

Chapter Title

This title is available via Amazon and other book sellers

Companion Book By This Author

The Cutting Edge of AI Autonomous Cars

by Dr. Lance B. Eliot, MBA, PhD

This title is available via Amazon and other book sellers

ABOUT THE AUTHOR

Dr. Lance B. Eliot, MBA, PhD is the CEO of Techbruim, Inc. and Executive Director of the Cybernetic Self-Driving Car Institute, and has over twenty years of industry experience including serving as a corporate officer in a billion dollar firm and was a partner in a major executive services firm. He is also a serial entrepreneur having founded, ran, and sold several high-tech related businesses. He previously hosted the popular radio show *Technotrends* that was also available on American Airlines flights via their in-flight audio program. Author or co-author of a dozen books and over 400 articles, he has made appearances on CNN, and has been a frequent speaker at industry conferences.

A former professor at the University of Southern California (USC), he founded and led an innovative research lab on Artificial Intelligence in Business. Known as the "AI Insider" his writings on AI advances and trends has been widely read and cited. He also previously served on the faculty of the University of California Los Angeles (UCLA), and was a visiting professor at other major universities. He was elected to the International Board of the Society for Information Management (SIM), a prestigious association of over 3,000 high-tech executives worldwide.

He has performed extensive community service, including serving as Senior Science Adviser to the Vice Chair of the Congressional Committee on Science & Technology. He has served on the Board of the OC Science & Engineering Fair (OCSEF), where he is also has been a Grand Sweepstakes judge, and likewise served as a judge for the Intel International SEF (ISEF). He served as the Vice Chair of the Association for Computing Machinery (ACM) Chapter, a prestigious association of computer scientists. Dr. Eliot has been a shark tank judge for the USC Mark Stevens Center for Innovation on start-up pitch competitions, and served as a mentor for several incubators and accelerators in Silicon Valley and Silicon Beach. He served on several Boards and Committees at USC, including having served on the Marshall Alumni Association (MAA) Board in Southern California.

Dr. Eliot holds a PhD from USC, MBA, and Bachelor's in Computer Science, and earned the CDP, CCP, CSP, CDE, and CISA certifications. Born and raised in Southern California, and having traveled and lived internationally, he enjoys scuba diving, surfing, and sailing.

ADDENDUM

The Cutting Edge of AI Autonomous Cars

Practical Advances in Artificial Intelligence (AI) and Machine Learning

By
Dr. Lance B. Eliot, MBA, PhD

For supplemental materials of this book, visit:

www.ai-selfdriving-cars.guru

For special orders of this book, contact:

LBE Press Publishing

Email: LBE.Press.Publishing@gmail.com

www.ingramcontent.com/pod-product-compliance
Lightning Source LLC
Chambersburg PA
CBHW021553210326
41599CB00010B/426